精通 Python 网络编程
第 2 版(影印版)
Mastering Python Networking,
2nd Edition

Eric Chou 著

南京　东南大学出版社

图书在版编目(CIP)数据

精通 Python 网络编程:英文/(美)埃里克·周(Eric Chou)著.—2 版(影印本).—南京:东南大学出版社,2019.3

书名原文:Mastering Python Networking, 2nd Edition
ISBN 978-7-5641-8294-6

Ⅰ.①精… Ⅱ.①埃… Ⅲ.①软件工具—程序设计—英文 Ⅳ.①TP311.561

中国版本图书馆 CIP 数据核字(2019)第 025732 号
图字:10-2018-494 号

© 2018 by PACKT Publishing Ltd.

Reprint of the English Edition, jointly published by PACKT Publishing Ltd and Southeast University Press, 2019. Authorized reprint of the original English edition, 2018 PACKT Publishing Ltd, the owner of all rights to publish and sell the same.

All rights reserved including the rights of reproduction in whole or in part in any form.

英文原版由 PACKT Publishing Ltd 出版 2018。

英文影印版由东南大学出版社出版 2019。此影印版的出版和销售得到出版权和销售权的所有者 —— PACKT Publishing Ltd 的许可。

版权所有,未得书面许可,本书的任何部分和全部不得以任何形式重制。

精通 Python 网络编程 第 2 版(影印版)

出版发行:东南大学出版社
地　　址:南京四牌楼 2 号　邮编:210096
出 版 人:江建中
网　　址:http://www.seupress.com
电子邮件:press@seupress.com
印　　刷:常州市武进第三印刷有限公司
开　　本:787 毫米×980 毫米　16 开本
印　　张:29
字　　数:568 千字
版　　次:2019 年 3 月第 1 版
印　　次:2019 年 3 月第 1 次印刷
书　　号:ISBN 978-7-5641-8294-6
定　　价:106.00 元

本社图书若有印装质量问题,请直接与营销部联系。电话(传真):025-83791830

mapt.io

Mapt is an online digital library that gives you full access to over 5,000 books and videos, as well as industry leading tools to help you plan your personal development and advance your career. For more information, please visit our website.

Why subscribe?

- Spend less time learning and more time coding with practical eBooks and Videos from over 4,000 industry professionals
- Improve your learning with Skill Plans built especially for you
- Get a free eBook or video every month
- Mapt is fully searchable
- Copy and paste, print, and bookmark content

PacktPub.com

Did you know that Packt offers eBook versions of every book published, with PDF and ePub files available? You can upgrade to the eBook version at www.PacktPub.com and as a print book customer, you are entitled to a discount on the eBook copy. Get in touch with us at service@packtpub.com for more details.

At www.PacktPub.com, you can also read a collection of free technical articles, sign up for a range of free newsletters, and receive exclusive discounts and offers on Packt books and eBooks.

Contributors

About the author

Eric Chou is a seasoned technologist with over 18 years of industry experience. He has worked on and helped managed some of the largest networks in the industry while working at Amazon AWS, Microsoft Azure, and other companies. Eric is passionate about network automation, Python, and helping companies build better security postures. Eric is the author of several books and online classes on networking with Python and network security. He is the proud inventor of two patents in IP telephony. Eric shares his deep interest in technology through his books, classes, and his blog, and contributes to some of the popular Python open source projects.

> *I would like to thank the open source and Python community members for generously sharing their knowledge and code with the public. Without their contribution, many of the projects referenced in this book would not have been possible.*
>
> *I would like to thank the Packt Publishing team for the opportunity to work on the second edition of the book, and the technical reviewer, Rickard Körkkö, for generously agreeing to review the book.*
>
> *To my wife and best friend, Joanna, I won the lottery the day I met you. To my two girls, Mikaelyn and Esmie, you make me so proud, I love you both dearly.*

About the reviewer

Rickard Körkkö, CCNP (Routing and Switching) and Cisco Network Programmability Design and Implementation Specialist, is a NetOps consultant at SDNit, where he's part of a group of experienced technical specialists with a great interest in and focus on emerging network technologies. His daily work includes working with orchestration tools such as Ansible to manage network devices. He's a self-taught programmer with a primary focus on Python. He has also served as a technical reviewer for the book *A Practical Guide to Linux Commands, Editors, and Shell Programming, Third Edition* by Mark G. Sobell.

Packt is searching for authors like you

If you're interested in becoming an author for Packt, please visit `authors.packtpub.com` and apply today. We have worked with thousands of developers and tech professionals, just like you, to help them share their insight with the global tech community. You can make a general application, apply for a specific hot topic that we are recruiting an author for, or submit your own idea.

Table of Contents

Preface — 1

Chapter 1: Review of TCP/IP Protocol Suite and Python — 7
  An overview of the internet — 8
    Servers, hosts, and network components — 9
    The rise of data centers — 10
      Enterprise data centers — 10
      Cloud data centers — 11
      Edge data centers — 12
  The OSI model — 13
  Client-server model — 15
  Network protocol suites — 15
    The transmission control protocol — 16
      Functions and characteristics of TCP — 16
      TCP messages and data transfer — 17
    User datagram protocol — 17
    The internet protocol — 18
      The IP NAT and security — 19
      IP routing concepts — 20
  Python language overview — 20
    Python versions — 22
    Operating system — 23
    Running a Python program — 23
    Python built-in types — 24
      The None type — 25
      Numerics — 25
      Sequences — 25
      Mapping — 28
      Sets — 29
    Python operators — 29
    Python control flow tools — 30
    Python functions — 32
    Python classes — 33
    Python modules and packages — 34
  Summary — 36

Chapter 2: Low-Level Network Device Interactions — 37
  The challenges of the CLI — 38
  Constructing a virtual lab — 39
    Cisco VIRL — 41
      VIRL tips — 43

Table of Contents

Cisco DevNet and dCloud	46
GNS3	48
Python Pexpect library	**49**
Pexpect installation	49
Pexpect overview	50
Our first Pexpect program	55
More Pexpect features	56
Pexpect and SSH	57
Putting things together for Pexpect	58
The Python Paramiko library	**59**
Installation of Paramiko	60
Paramiko overview	61
Our first Paramiko program	64
More Paramiko features	65
Paramiko for servers	65
Putting things together for Paramiko	66
Looking ahead	**68**
Downsides of Pexpect and Paramiko compared to other tools	68
Idempotent network device interaction	68
Bad automation speeds bad things up	69
Summary	**69**
Chapter 3: APIs and Intent-Driven Networking	**71**
Infrastructure as code	**72**
Intent-Driven Networking	73
Screen scraping versus API structured output	74
Data modeling for infrastructure as code	77
The Cisco API and ACI	**78**
Cisco NX-API	79
Lab software installation and device preparation	79
NX-API examples	80
The Cisco and YANG models	85
The Cisco ACI	86
The Python API for Juniper networks	**88**
Juniper and NETCONF	89
Device preparation	89
Juniper NETCONF examples	91
Juniper PyEZ for developers	93
Installation and preparation	94
PyEZ examples	96
The Arista Python API	**98**
Arista eAPI management	98
The eAPI preparation	99
eAPI examples	102
The Arista Pyeapi library	104
Pyeapi installation	104
Pyeapi examples	105

Vendor-neutral libraries — 109
Summary — 109

Chapter 4: The Python Automation Framework – Ansible Basics — 111
A more declarative framework — 113
A quick Ansible example — 114
    The control node installation — 115
    Running different versions of Ansible from source — 116
    Lab setup — 117
    Your first Ansible playbook — 118
        The public key authorization — 118
        The inventory file — 119
        Our first playbook — 120
The advantages of Ansible — 122
    Agentless — 122
    Idempotent — 123
    Simple and extensible — 124
    Network vendor support — 125
The Ansible architecture — 126
    YAML — 127
    Inventories — 128
    Variables — 129
    Templates with Jinja2 — 133
Ansible networking modules — 133
    Local connections and facts — 133
    Provider arguments — 134
The Ansible Cisco example — 136
    Ansible 2.5 connection example — 139
The Ansible Juniper example — 141
The Ansible Arista example — 143
Summary — 145

Chapter 5: The Python Automation Framework – Beyond Basics — 147
Ansible conditionals — 148
    The when clause — 148
    Ansible network facts — 151
    Network module conditional — 153
Ansible loops — 155
    Standard loops — 155
    Looping over dictionaries — 156
Templates — 158
    The Jinja2 template — 160
    Jinja2 loops — 161
    The Jinja2 conditional — 162
Group and host variables — 164

Group variables	165
Host variables	166
The Ansible Vault	**167**
The Ansible include and roles	**169**
The Ansible include statement	169
Ansible roles	171
Writing your own custom module	**174**
The first custom module	175
The second custom module	177
Summary	**179**

Chapter 6: Network Security with Python — 181
The lab setup	**182**
Python Scapy	**185**
Installing Scapy	186
Interactive examples	187
Sniffing	188
The TCP port scan	190
The ping collection	193
Common attacks	194
Scapy resources	195
Access lists	**196**
Implementing access lists with Ansible	197
MAC access lists	200
The Syslog search	**201**
Searching with the RE module	202
Other tools	**204**
Private VLANs	204
UFW with Python	205
Further reading	206
Summary	**207**

Chapter 7: Network Monitoring with Python – Part 1 — 209
Lab setup	**210**
SNMP	**211**
Setup	212
PySNMP	215
Python for data visualization	**221**
Matplotlib	222
Installation	222
Matplotlib – the first example	222
Matplotlib for SNMP results	225
Additional Matplotlib resources	230
Pygal	230
Installation	230
Pygal – the first example	231

Pygal for SNMP results	233
Additional Pygal resources	236
Python for Cacti	236
Installation	237
Python script as an input source	239
Summary	241

Chapter 8: Network Monitoring with Python – Part 2 — 243
Graphviz — 245
- Lab setup — 246
- Installation — 247
- Graphviz examples — 248
- Python with Graphviz examples — 250
- LLDP neighbor graphing — 251
  - Information retrieval — 253
  - Python parser script — 254
  - Final playbook — 258

Flow-based monitoring — 260
- NetFlow parsing with Python — 261
  - Python socket and struct — 263
- ntop traffic monitoring — 266
  - Python extension for ntop — 270
- sFlow — 274
  - SFlowtool and sFlow-RT with Python — 274

Elasticsearch (ELK stack) — 278
- Setting up a hosted ELK service — 279
- The Logstash format — 280
- Python helper script for Logstash formatting — 281

Summary — 283

Chapter 9: Building Network Web Services with Python — 285
Comparing Python web frameworks — 287
Flask and lab setup — 289
Introduction to Flask — 291
- The HTTPie client — 293
- URL routing — 294
- URL variables — 295
- URL generation — 296
- The jsonify return — 297

Network resource API — 298
- Flask-SQLAlchemy — 298
- Network content API — 300
- Devices API — 303
- The device ID API — 305

Network dynamic operations — 306
- Asynchronous operations — 309

Security	311
Additional resources	314
Summary	315
Chapter 10: AWS Cloud Networking	**317**
AWS setup	318
AWS CLI and Python SDK	321
AWS network overview	324
Virtual private cloud	328
Route tables and route targets	333
Automation with CloudFormation	335
Security groups and the network ACL	339
Elastic IP	341
NAT Gateway	342
Direct Connect and VPN	344
VPN Gateway	344
Direct Connect	345
Network scaling services	347
Elastic Load Balancing	347
Route53 DNS service	348
CloudFront CDN services	348
Other AWS network services	349
Summary	349
Chapter 11: Working with Git	**351**
Introduction to Git	352
Benefits of Git	353
Git terminology	354
Git and GitHub	354
Setting up Git	355
Gitignore	356
Git usage examples	357
GitHub example	364
Collaborating with pull requests	368
Git with Python	372
GitPython	372
PyGitHub	373
Automating configuration backup	375
Collaborating with Git	378
Summary	379
Chapter 12: Continuous Integration with Jenkins	**381**
Traditional change-management process	381
Introduction to continuous integration	383
Installing Jenkins	384

Jenkins example — 387
First job for the Python script — 388
Jenkins plugins — 394
Network continuous integration example — 396
Jenkins with Python — 405
Continuous integration for Networking — 407
Summary — 407

Chapter 13: Test-Driven Development for Networks — 409
Test-driven development overview — 410
Test definitions — 411
Topology as code — 411
Python's unittest module — 416
More on Python testing — 420
pytest examples — 420
Writing tests for networking — 423
Testing for reachability — 423
Testing for network latency — 425
Testing for security — 426
Testing for transactions — 426
Testing for network configuration — 427
Testing for Ansible — 427
Pytest in Jenkins — 428
Jenkins integration — 428
Summary — 433

Other Books You May Enjoy — 435

Index — 441

Preface

As Charles Dickens wrote in *A Tale of Two Cities*, "It was the best of times, it was the worse of times, it was the age of wisdom, it was the age of foolishness." His seemingly contradictory words perfectly describe the chaos and mood felt during a time of change and transition. We are no doubt experiencing a similar time with the rapid changes in the fields of network engineering. As software development becomes more integrated into all aspects of networking, the traditional command-line interface and vertically integrated network stack methods are no longer the best ways to manage today's networks. For network engineers, the changes we are seeing are full of excitement and opportunities and yet challenging, particularly for those who need to quickly adapt and keep up. This book has been written to help ease the transition for networking professionals by providing a practical guide that addresses how to evolve from a traditional platform to one built on software-driven practices.

In this book, we use Python as the programming language of choice to master network engineering tasks. Python is an easy-to-learn, high-level programming language that can effectively complement network engineers' creativity and problem-solving skills to streamline daily operations. Python is becoming an integral part of many large-scale networks, and through this book, I hope to share with you the lessons I've learned.

Since the publication of the first edition, I have been able to have interesting and meaningful conversations with many of the readers of the book. I am humbled by the success of the first edition of the book and took to the heart of the feedback I was given. In the second edition, I have tried to make the examples and technologies more relevant. In particular, the traditional OpenFlow SDN chapters were replaced with some of the Network DevOps tools. I sincerely hope the new addition is useful to you.

A time of change presents great opportunities for technological advancement. The concepts and tools in this book have helped me tremendously in my career, and I hope they can do the same for you.

Who this book is for

This book is ideal for IT professionals and operations engineers who already manage groups of network devices and would like to expand their knowledge on using Python and other tools to overcome network challenges. Basic knowledge of networking and Python is recommended.

What this book covers

Chapter 1, *Review of TCP/IP Protocol Suite and Python*, reviews the fundamental technologies that make up internet communication today, from the OSI and client-server model to TCP, UDP, and IP protocol suites. The chapter will review the basics of Python languages such as types, operators, loops, functions, and packages.

Chapter 2, *Low-Level Network Device Interactions*, uses practical examples to illustrate how to use Python to execute commands on a network device. It will also discuss the challenges of having a CLI-only interface in automation. The chapter will use the Pexpect and Paramiko libraries for the examples.

Chapter 3, *APIs and Intent-Driven Networking*, discusses the newer network devices that support **Application Programming Interfaces** (**APIs**) and other high-level interaction methods. It also illustrates tools that allow abstraction of low-level tasks while focusing on the intent of the network engineers. A discussion about and examples of Cisco NX-API, Juniper PyEZ, and Arista Pyeapi will be used in the chapter.

Chapter 4, *The Python Automation Framework – Ansible Basics*, discusses the basics of Ansible, an open source, Python-based automation framework. Ansible moves one step further from APIs and focuses on declarative task intent. In this chapter, we will cover the advantages of using Ansible, its high-level architecture, and see some practical examples of Ansible with Cisco, Juniper, and Arista devices.

Chapter 5, *The Python Automation Framework – Beyond Basics*, builds on the knowledge in the previous chapter and covers the more advanced Ansible topics. We will cover conditionals, loops, templates, variables, Ansible Vault, and roles. It will also cover the basics of writing custom modules.

Chapter 6, *Network Security with Python*, introduces several Python tools to help you secure your network. It will discuss using Scapy for security testing, using Ansible to quickly implement access lists, and using Python for network forensic analysis.

Chapter 7, *Network Monitoring with Python – Part 1*, covers monitoring the network using various tools. The chapter contains some examples using SNMP and PySNMP for queries to obtain device information. Matplotlib and Pygal examples will be shown for graphing the results. The chapter will end with a Cacti example using a Python script as an input source.

Chapter 8, *Network Monitoring with Python – Part 2*, covers more network monitoring tools. The chapter will start with using Graphviz to graph the network from LLDP information. We will move to use examples with push-based network monitoring using Netflow and other technologies. We will use Python to decode flow packets and ntop to visualize the results. An overview of Elasticsearch and how it can be used for network monitoring will also be covered.

Chapter 9, *Building Network Web Services with Python*, shows you how to use the Python Flask web framework to create our own API for network automation. The network API offers benefits such as abstracting the requester from network details, consolidating and customizing operations, and providing better security by limiting the exposure of available operations.

Chapter 10, *AWS Cloud Networking*, shows how we can use AWS to build a virtual network that is functional and resilient. We will cover virtual private cloud technologies such as CloudFormation, VPC routing table, access-list, Elastic IP, NAT Gateway, Direct Connect, and other related topics.

Chapter 11, *Working with Git*, we will illustrate how we can leverage Git for collaboration and code version control. Practical examples of using Git for network operations will be used in this chapter.

Chapter 12, *Continuous Integration with Jenkins*, uses Jenkins to automatically create operations pipelines that can save us time and increase reliability.

Chapter 13, *Test-Driven Development for Networks*, explains how to use Python's unittest and PyTest to create simple tests to verify our code. We will also see examples of writing tests for our network to verify reachability, network latency, security, and network transactions. We will also see how we can integrate the tests into continuous integration tools, such as Jenkins.

To get the most out of this book

To get the most out of this book, some basic hands-on network operation knowledge and Python is recommended. Most of the chapters can be read in any order, with the exceptions of chapters 4 and 5, which should be read in sequence. Besides the basic software and hardware tools introduced at the beginning of the book, new tools relevant to each of the chapters will be introduced.

It is highly recommended to follow and practice the examples shown in your own network lab.

Download the example code files

You can download the example code files for this book from your account at www.packtpub.com. If you purchased this book elsewhere, you can visit www.packtpub.com/support and register to have the files emailed directly to you.

You can download the code files by following these steps:

1. Log in or register at www.packtpub.com.
2. Select the **SUPPORT** tab.
3. Click on **Code Downloads & Errata**.
4. Enter the name of the book in the **Search** box and follow the onscreen instructions.

Once the file is downloaded, please make sure that you unzip or extract the folder using the latest version of:

- WinRAR/7-Zip for Windows
- Zipeg/iZip/UnRarX for Mac
- 7-Zip/PeaZip for Linux

The code bundle for the book is also hosted on GitHub at https://github.com/PacktPublishing/Mastering-Python-Networking-Second-Edition. In case there's an update to the code, it will be updated on the existing GitHub repository.

We also have other code bundles from our rich catalog of books and videos available at https://github.com/PacktPublishing/. Check them out!

Download the color images

We also provide a PDF file that has color images of the screenshots/diagrams used in this book. You can download it here: https://www.packtpub.com/sites/default/files/downloads/MasteringPythonNetworkingSecondEdition_ColorImages.pdf.

Conventions used

There are a number of text conventions used throughout this book.

`CodeInText`: Indicates code words in text, database table names, folder names, filenames, file extensions, pathnames, dummy URLs, user input, and Twitter handles. Here is an example: "The auto-config also generated `vty` access for both telnet and SSH."

A block of code is set as follows:

```
# This is a comment
print("hello world")
```

Any command-line input or output is written as follows:

```
$ python
Python 2.7.12 (default, Dec 4 2017, 14:50:18)
[GCC 5.4.0 20160609] on linux2
Type "help", "copyright", "credits" or "license" for more information.
>>> exit()
```

Bold: Indicates a new term, an important word, or words that you see onscreen. For example, words in menus or dialog boxes appear in the text like this. Here is an example: "In the **Topology Design** option, I set the **Management Network** option to **Shared Flat Network** in order to use VMnet2 as the management network on the virtual routers."

Warnings or important notes appear like this.

Tips and tricks appear like this.

Get in touch

Feedback from our readers is always welcome.

General feedback: Email `feedback@packtpub.com` and mention the book title in the subject of your message. If you have questions about any aspect of this book, please email us at `questions@packtpub.com`.

Errata: Although we have taken every care to ensure the accuracy of our content, mistakes do happen. If you have found a mistake in this book, we would be grateful if you would report this to us. Please visit `www.packtpub.com/submit-errata`, selecting your book, clicking on the Errata Submission Form link, and entering the details.

Piracy: If you come across any illegal copies of our works in any form on the Internet, we would be grateful if you would provide us with the location address or website name. Please contact us at `copyright@packtpub.com` with a link to the material.

If you are interested in becoming an author: If there is a topic that you have expertise in and you are interested in either writing or contributing to a book, please visit `authors.packtpub.com`.

Reviews

Please leave a review. Once you have read and used this book, why not leave a review on the site that you purchased it from? Potential readers can then see and use your unbiased opinion to make purchase decisions, we at Packt can understand what you think about our products, and our authors can see your feedback on their book. Thank you!

For more information about Packt, please visit `packtpub.com`.

Review of TCP/IP Protocol Suite and Python

Welcome to the new age of network engineering. When I first started working as a network engineer 18 years ago, at the turn of the millennium, the role was distinctly different than other technical roles. Network engineers mainly possessed domain-specific knowledge to manage and operate local and wide area networks, while occasionally crossing over to systems administration, but there was no expectation to write code or understand programming concepts. This is no longer the case. Over the years, the DevOps and **Software-Defined Networking (SDN)** movement, among other factors, have significantly blurred the lines between network engineers, systems engineers, and developers.

The fact that you have picked up this book suggests that you might already be an adopter of network DevOps, or maybe you are considering going down that path. Maybe you have been working as a network engineer for years, just as I was, and want to know what the buzz around the Python programming language is about. Or you might already be fluent in Python but wonder what its applications are in network engineering. If you fall into any of these camps, or are simply just curious about Python in the network engineering field, I believe this book is for you:

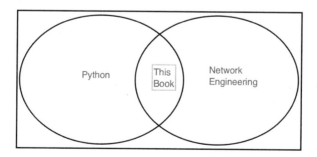

The intersection between Python and network engineering

Many books that dive into the topics of network engineering and Python have already been written. I do not intend to repeat their efforts with this book. Instead, this book assumes that you have some hands-on experience of managing networks, as well as a basic understanding of network protocols and the Python language. You do not need to be an expert in Python or network engineering, but should find that the concepts in this chapter form a general review. The rest of the chapter should set the level of expectation of the prior knowledge required to get the most out of this book. If you want to brush up on the contents of this chapter, there are lots of free or low-cost resources to bring you up to speed. I would recommend the free Khan Academy (https://www.khanacademy.org/) and the Python tutorials at: https://www.python.org/.

This chapter will pay a very quick visit to the relevant networking topics. From my experience working in the field, a typical network engineer or developer might not remember the exact TCP state machine to accomplish their daily tasks (I know I don't), but they would be familiar with the basics of the OSI model, the TCP and UDP operations, different IP headers fields, and other fundamental concepts.

We will also look at a high-level review of the Python language; just enough for those readers who do not code in Python on a daily basis to have ground to walk on for the rest of the book.

Specifically, we will cover the following topics:

- An overview of the internet
- The OSI and client-server model
- TCP, UDP, and IP protocol suites
- Python syntax, types, operators, and loops
- Extending Python with functions, classes, and packages

Of course, the information presented in this chapter is not exhaustive; please do check out the references for further information.

An overview of the internet

What is the internet? This seemingly easy question might receive different answers depending on your background. The internet means different things to different people; the young, the old, students, teachers, business people, poets, could all give different answers to the question.

To a network engineer, the internet is a global computer network consisting of a web of inter-networks connecting large and small networks together. In other words, it is a network of networks without a centralized owner. Take your home network as an example. It might consist of a home Ethernet switch and a wireless access point connecting your smartphone, tablet, computers, and TV together for the devices to communicate with each other. This is your **Local Area Network (LAN)**. When your home network needs to communicate with the outside world, it passes information from your LAN to a larger network, often appropriately named the **Internet Service Provider (ISP)**. Your ISP often consists of edge nodes that aggregate the traffic to their core network. The core network's function is to interconnect these edge networks via a higher speed network. At special edge nodes, your ISP is connected to other ISPs to pass your traffic appropriately to your destination. The return path from your destination to your home computer, tablet, or smartphone may or may not follow the same path through all of these networks back to your device, while the source and destination remain the same.

Let's take a look at the components making up this web of networks.

Servers, hosts, and network components

Hosts are end nodes on the network that communicate to other nodes. In today's world, a host can be a traditional computer, or can be your smartphone, tablet, or TV. With the rise of the **Internet of Things (IoT)**, the broad definition of a host can be expanded to include an IP camera, TV set-top boxes, and the ever-increasing type of sensors that we use in agriculture, farming, automobiles, and more. With the explosion of the number of hosts connected to the internet, all of them need to be addressed, routed, and managed. The demand for proper networking has never been greater.

Most of the time when we are on the internet, we make requests for services. This could be viewing a web page, sending or receiving emails, transferring files, and so on. These services are provided by **servers**. As the name implies, servers provide services to multiple nodes and generally have higher levels of hardware specification because of this. In a way, servers are special super-nodes on the network that provide additional capabilities to its peers. We will look at servers later on in the client-server model section.

If you think of servers and hosts as cities and towns, the **network components** are the roads and highways that connect them together. In fact, the term information superhighway comes to mind when describing the network components that transmit the ever increasing bits and bytes across the globe. In the OSI model that we will look at in a bit, these network components are layer one to three devices. They are layer two and three routers and switches that direct traffic, as well as layer one transports such as fiber optic cables, coaxial cables, twisted copper pairs, and some DWDM equipment, to name a few.

Collectively, hosts, servers, and network components make up the internet as we know it today.

The rise of data centers

In the last section, we looked at the different roles that servers, hosts, and network components play in the inter-network. Because of the higher hardware capacity that servers demand, they are often put together in a central location, so they can be managed more efficiently. We often refer to these locations as data centers.

Enterprise data centers

In a typical enterprise, the company generally has the need for internal tools such as emailing, document storage, sales tracking, ordering, HR tools, and a knowledge sharing intranet. These services translate into file and mail servers, database servers, and web servers. Unlike user computers, these are generally high-end computers that require a lot of power, cooling, and network connections. A byproduct of the hardware is also the amount of noise they make. They are generally placed in a central location, called the **Main Distribution Frame** (**MDF**), in the enterprise to provide the necessary power feed, power redundancy, cooling, and network connectivity.

To connect to the MDF, the user's traffic is generally aggregated at a location closer to the user, which is sometimes called the **Intermediate Distribution Frame** (**IDF**), before they are bundled up and connected to the MDF. It is not unusual for the IDF-MDF spread to follow the physical layout of the enterprise building or campus. For example, each building floor can consist of an IDF that aggregates to the MDF on another floor. If the enterprise consists of several buildings, further aggregation can be done by combining the buildings' traffic before connecting them to the enterprise data center.

Enterprise data centers generally follow the network design of three layers. These layers are access, distribution, and a core. The access layer is analogous to the ports each user connects to, the IDF can be thought of as the distribution layer, while the core layer consists of the connection to the MDF and the enterprise data centers. This is, of course, a generalization of enterprise networks, as some of them will not follow the same model.

Chapter 1

Cloud data centers

With the rise of cloud computing and software, or infrastructure as a service, the data centers cloud providers build are at a hyper-scale. Because of the number of servers they house, they generally demand a much, much higher capacity for power, cooling, network speed, and feed than any enterprise data center. Even after working on cloud provider data centers for many years, every time I visit a cloud provider data center, I am still amazed at the scale of them. In fact, cloud data centers are so big and power-hungry that they are typically built close to power plants where they can get the cheapest power rate, without losing too much efficiency during the transportation of the power. Their cooling needs are so great, some are forced to be creative about where the data center is built, building in a generally cold climate just so they can just open the doors and windows to keep the server running at a safe temperature when needed. Any search engine can give you some of the astounding numbers when it comes to the science of building and managing cloud data centers for the likes of Amazon, Microsoft, Google, and Facebook:

Utah data center (source: https://en.wikipedia.org/wiki/Utah_Data_Center)

At the cloud provider scale, the services that they need to provide are generally not cost efficient or able to feasibly be housed in a single server. They are spread between a fleet of servers, sometimes across many different racks, to provide redundancy and flexibility for service owners. The latency and redundancy requirements put a tremendous amount of pressure on the network. The number of interconnections equates to an explosive growth of network equipment; this translates into the number of times this network equipment needs to be racked, provisioned, and managed. A typical network design would be a multi-staged, CLOS network:

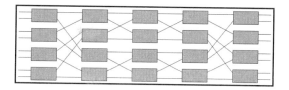

CLOS network

[11]

In a way, cloud data centers are where network automation becomes a necessity for speed and reliability. If we follow the traditional way of managing network devices via a Terminal and command-line interface, the number of engineering hours required would not allow the service to be available in a reasonable amount of time. This is not to mention that human repetition is error-prone, inefficient, and a terrible waste of engineering talent.

Cloud data centers are where I started my path of network automation with Python a number of years ago, and I've never looked back since.

Edge data centers

If we have sufficient computing power at the data center level, why keep anything anywhere else but at these data centers? All the connections from clients around the world can be routed back to the data center servers providing the service, and we can call it a day, right? The answer, of course, depends on the use case. The biggest limitation in routing the request and session all the way back from the client to a large data center is the latency introduced in the transport. In other words, large latency is where the network becomes a bottleneck. The latency number would never be zero: even as fast as light can travel in a vacuum, it still takes time for physical transportation. In the real world, latency would be much higher than light in a vacuum when the packet is traversing through multiple networks, and sometimes through an undersea cable, slow satellite links, 3G or 4G cellular links, or Wi-Fi connections.

The solution? Reduce the number of networks the end user traverses through. Be as closely connected to the user as possible at the edge where the user enters your network and place enough resources at the edge location to serve the request. Let's take a minute and imagine that you are building the next generation of video streaming service. In order to increase customer satisfaction with smooth streaming, you would want to place the video server as close to the customer as possible, either inside or very near to the customer's ISP. Also, the upstream of the video server farm would not just be connected to one or two ISPs, but all the ISPs that I can connect to to reduce the hop count. All the connections would be with as much bandwidth as needed to decrease latency during peak hours. This need gave rise to the peering exchange's edge data centers of big ISP and content providers. Even when the number of network devices is not as high as cloud data centers, they too can benefit from network automation in terms of the increased reliability, security, and visibility network automation brings.

We will cover security and visibility in later chapters of this book.

The OSI model

No network book is complete without first going over the **Open System Interconnection (OSI)** model. The model is a conceptional model that componentizes the telecommunication functions into different layers. The model defines seven layers, and each layer sits independently on top of another one, as long as they follow defined structures and characteristics. For example, in the network layer, IP, can sit on top of the different types of data link layers, such as Ethernet or frame relay. The OSI reference model is a good way to normalize different and diverse technologies into a set of common language that people can agree on. This greatly reduces the scope for parties working on individual layers and allows them to look at specific tasks in depth without worrying too much about compatibility:

	Layer	Protocol data unit (PDU)	Function
Host Layers	7. Application	Data	High-level APIs, including resource sharing, remote file access
	6. Presentation		Translation of data between a networking service and an application; including character encoding, data compression and encryption / decryption
	5. Session		Managing communication sessions, i.e. continuous exchange of information in the form of mutiple back-and-forth transmissions between two nodes
	4. Transport	Segment (TCP) /Datagram (UDP)	Reliable transmission of data segments between points on a network, including segmentation, acknowledgement and multiplexing
Media Layers	3. Network	Packet	Structuring and managing a multi-node network, including addressing, routing and traffic control
	2. Data link	Frame	Reliable transmission of data frames between two nodes connected by a physical layer
	1. Physical	Bit	Transmission and reception of raw bit streams over a physical medium

OSI model

The OSI model was initially worked on in the late 1970s and was later published jointly by the **International Organization for Standardization (ISO)** and what's now known as the **Telecommunication Standardization Sector** of the **International Telecommunication Union (ITU-T)**. It is widely accepted and commonly referred to when introducing a new topic in telecommunication.

Review of TCP/IP Protocol Suite and Python

Around the same time period of the OSI model development, the internet was taking shape. The reference model the original designer used is often referred to as the TCP/IP model. The **Transmission Control Protocol (TCP)** and the **Internet Protocol (IP)** were the original protocol suites contained in the design. This is somewhat similar to the OSI model in the sense that they divide end-to-end data communication into abstraction layers. What is different is the model combines layers 5 to 7 in the OSI model in the **Application** layer, while the **Physical** and **Data link** layers are combined in the **Link** layer:

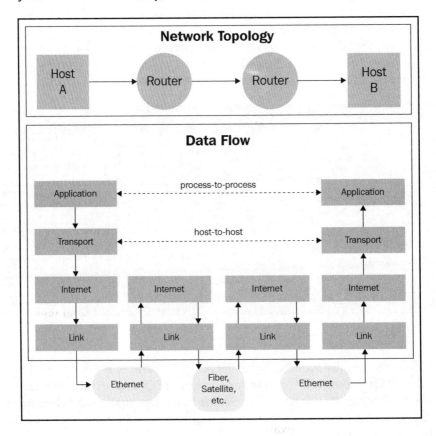

Internet protocol suite

Both the OSI and TCP/IP models are useful for providing standards for end-to-end data communication. However, for the most part, we will refer to the TCP/IP model more, since that is what the internet was built on. We will specify the OSI model when needed, such as when we are discussing the web framework in upcoming chapters.

Client-server model

The reference models demonstrated a standard way for data to communicate between two nodes. Of course, by now, we all know that not all nodes are created equal. Even in its DARPA-net days, there were workstation nodes, and there were nodes with the purpose of providing content to other nodes. These server nodes typically have higher hardware specifications and are managed more closely by engineers. Since these nodes provide resources and services to others, they are typically referred to as servers. Servers typically sit idle, waiting for clients to initiate requests for their resources. This model of distributed resources that are asked for by the client is referred to as the client-server model.

Why is this important? If you think about it for a minute, the importance of networking is highlighted by this client-server model. Without it, there is really not a lot of need for network interconnections. It is the need to transfer bits and bytes from client to server that shines a light on the importance of network engineering. Of course, we are all aware of how the biggest network of them all, the internet, has been transforming the lives of all of us and continuing to do so.

How, you asked, can each node determine the time, speed, source, and destination every time they need to talk to each other? This brings us to network protocols.

Network protocol suites

In the early days of computer networking, protocols were proprietary and closely controlled by the company who designed the connection method. If you were using **Novell's IPX/SPX** protocol in your hosts, you would not able to communicate with Apple's **AppleTalk** hosts and vice versa. These proprietary protocol suites generally have analogous layers to the OSI reference model and follow the client-server communication method. They generally work great in **Local Area Networks** (**LAN**) that are closed, without the need to communicate with the outside world. When traffic does need to move beyond the local LAN, typically, an internet working device, such as a router, is used to translate from one protocol to another. An example would be a router connecting an AppleTalk network to an IP-based network. The translation is usually not perfect, but since most of the communication happens within the LAN in the early days, it is okay.

However, as the need for inter-network communication rises beyond the LAN, the need for standardizing the network protocol suites becomes greater. The proprietary protocols eventually gave way to the standardized protocol suites of TCP, UDP, and IP, which greatly enhanced the ability of one network to talk to another. The internet, the greatest network of them all, relies on these protocols to function properly. In the next few sections, we will take a look at each of the protocol suites.

The transmission control protocol

The **Transmission Control Protocol** (**TCP**) is one of the main protocols used on the internet today. If you have opened a web page or have sent an email, you have come across the TCP protocol. The protocol sits at layer 4 of the OSI model, and it is responsible for delivering the data segment between two nodes in a reliable and error-checked manner. The TCP consists of a 160-bit header consisting of, among others, source and destination ports, a sequence number, an acknowledgment number, control flags, and a checksum:

Offsets	Octet	0															1									2								3							
Octet	Bit	0	1	2	3	4	5	6	7	8	9	10	11	12	13	14	15	16	17	18	19	20	21	22	23	24	25	26	27	28	29	30	31
0	0	Source port																	Destination port																
4	32	Sequence number																															
8	64	Acknowledgment number (if ACK set)																															
12	96	Data offset			Reserved 000			N S	C W R	E C E	U R G	A C K	P S H	R S T	S Y N	F I N	Window Size																
16	128	Checksum																	Urgent pointer (if URG set)																
20	160	Options (if data offset > 5. Padded at the end with "0" bytes if necessary.)																															
...	...	...																																	

TCP header

Functions and characteristics of TCP

TCP uses datagram sockets or ports to establish a host-to-host communication. The standard body, called **Internet Assigned Numbers Authority** (**IANA**) designates well-known ports to indicate certain services, such as port 80 for HTTP (web) and port 25 for SMTP (mail). The server in the client-server model typically listens on one of these well-known ports in order to receive communication requests from the client. The TCP connection is managed by the operating system by the socket that represents the local endpoint for connection.

The protocol operation consists of a state machine, where the machine needs to keep track of when it is listening for an incoming connection, during the communication session, as well as releasing resources once the connection is closed. Each TCP connection goes through a series of states such as `Listen`, `SYN-SENT`, `SYN-RECEIVED`, `ESTABLISHED`, `FIN-WAIT`, `CLOSE-WAIT`, `CLOSING`, `LAST-ACK`, `TIME-WAIT`, and `CLOSED`.

TCP messages and data transfer

The biggest difference between TCP and **User Datagram Protocol** (**UDP**), which is its close cousin on the same layer, is that it transmits data in an ordered and reliable fashion. The fact that the operation guarantees delivery is often referred to TCP as a connection-oriented protocol. It does this by first establishing a three-way handshake to synchronize the sequence number between the transmitter and the receiver, SYN, SYN-ACK, and ACK.

The acknowledgment is used to keep track of subsequent segments in the conversation. Finally, at the end of the conversation, one side will send a FIN message, and the other side will ACK the FIN message as well as sending a FIN message of its own. The FIN initiator will then ACK the FIN message that it received.

As many of us who have troubleshot a TCP connection can tell you, the operation can get quite complex. One can certainly appreciate that, most of the time, the operation just happens silently in the background.

A whole book could be written about the TCP protocol; in fact, many excellent books have been written on the protocol.

As this section is a quick overview, if interested, The TCP/IP Guide (http://www.tcpipguide.com/) is an excellent free resource that you can use to dig deeper into the subject.

User datagram protocol

The **User Datagram Protocol** (**UDP**) is also a core member of the internet protocol suite. Like TCP, it operates on layer 4 of the OSI model that is responsible for delivering data segments between the application and the IP layer. Unlike TCP, the header is only 64-bit, which only consists of a source and destination port, length, and checksum. The lightweight header makes it ideal for applications that prefer faster data delivery without setting up the session between two hosts or needing reliable data delivery. Perhaps it is hard to imagine with today's fast internet connections, but the extra header made a big difference to the speed of transmission in the early days of X.21 and frame relay links. Although, as important as the speed difference is, not having to maintain various states, such as TCP, also saves computer resources on the two endpoints:

Offsets	Octet	0								1								2								3						
Octet	Bit	0	1	2	3	4	5	6	7	8	9	10	11	12	13	14	15	16	17	18	19	20	21	22	23	24	25	26	27	28	29	30	31
0	0	Source port																Destination port															
4	32	Length																Checksum															

UDP header

You might now wonder why UDP was ever used at all in the modern age; given the lack of reliable transmission, wouldn't we want all the connections to be reliable and error-free? If you think about multimedia video streaming or Skype calling, those applications benefit from a lighter header when the application just wants to deliver the datagram as quickly as possible. You can also consider the fast DNS lookup process based on the UDP protocol. When the address you type in on the browser is translated into a computer understandable address, the user will benefit from a lightweight process, since this has to happen before even the first bit of information is delivered to you from your favorite website.

Again, this section does not do justice to the topic of UDP, and the reader is encouraged to explore the topic through various resources if you are is interested in learning more about UDP.

The internet protocol

As network engineers will tell you, they live at the **Internet Protocol** (IP) layer, which is layer 3 on the OSI model. **IP** has the job of addressing and routing between end nodes, among others. The addressing of an IP is probably its most important job. The address space is divided into two parts: the network and the host portion. The subnet mask is used to indicate which portion in the network address consists of the network and which portion is the host by matching the network portion with a 1 and the host portion with a 0. Both IPv4 and, later, IPv6 expresses the address in the dotted notation, for example, `192.168.0.1`. The subnet mask can either be in a dotted notation (`255.255.255.0`) or use a forward slash to express the number of bits that should be considered in the network bit (/24):

Offsets	Octet	0								1							2								3							
Octet	Bit	0	1	2	3	4	5	6	7	8	9	10	11	12	13	14	15	16	17	18	19	20	21	22	23	24	25	26	27	28	29	30	31
0	0	Version				IHL				DSCP						ECN		Total Length															
4	32	Identification																Flags			Fragment Offset												
8	64	Time To Live								Protocol								Header Checksum															
12	96	Source IP Address																															
16	128	Destination IP Address																															
20	160	Options (if IHL > 5)																															
24	192																																
28	224																																
32	256																																

IPv4 Header Format

IPv4 header

The IPv6 header, the next generation of the IP header of IPv4, has a fixed portion and various extension headers:

Offsets	Octet	0								1								2								3						
Octet	Bit	0	1	2	3	4	5	6	7	8	9	10	11	12	13	14	15	16	17	18	19	20	21	22	23	24	25	26	27	28	29	30	31
0	0	Version				Traffic Class												Flow Label															
4	32	Payload Length																Next Header								Hop Limit							
8	64	Source Address																														
12	96																																
16	128																																
20	160																																
24	192	Destination Address																														
28	224																																
32	256																																
36	288																																

IPv6 fixed header

The **Next Header** field in the fixed header section can indicate an extension header to be followed that carries additional information. The extension headers can include routing and fragment information. As much as the protocol designer would like to move from IPv4 to IPv6, the internet today is still pretty much addressed with IPv4, with some of the service provider networks addressed with IPv6 internally.

The IP NAT and security

Network Address Translation (**NAT**) is typically used for translating a range of private IPv4 addresses into publicly routable IPv4 addresses. But it can also mean a translation between IPv4 to IPv6, such as at a carrier edge when they use IPv6 inside of the network that needs to be translated to IPv4 when the packet leaves the network. Sometimes, NAT6 to 6 is used as well for security reasons.

Security is a continuous process that integrates all the aspects of networking, including automation and Python. This book aims to use Python to help you manage the network; security will be addressed as part of the following chapters in the book, such as using SSHv2 over telnet. We will also look at how we can use Python and other tools to gain visibility in the network.

IP routing concepts

In my opinion, IP routing is about having the intermediate devices between the two endpoints transmit the packets between them based on the IP header. For all communication via the internet, the packet will traverse through various intermediate devices. As mentioned, the intermediate devices consist of routers, switches, optical gears, and various other gears that do not examine beyond the network and transport layer. In a road trip analogy, you might travel in the United States from the city of San Diego in California to the city of Seattle in Washington. The IP source address is analogous to San Diego and the destination IP address can be thought of as Seattle. On your road trip, you will stop by many different intermediate spots, such as Los Angeles, San Francisco, and Portland; these can be thought of as the routers and switches between the source and destination.

Why was this important? In a way, this book is about managing and optimizing these intermediate devices. In the age of mega data centers that span the size of multiple American football fields, the need for efficient, agile, reliable, and cost-effective ways to manage the network becomes a major point of competitive advantage for companies. In future chapters, we will dive into how we can use Python programming to effectively manage a network.

Python language overview

In a nutshell, this book is about making our network engineering lives easier with Python. But what is Python and why is it the language of choice of many DevOps engineers? In the words of the Python Foundation **Executive Summary** (https://www.python.org/doc/essays/blurb/):

> "*Python is an interpreted, object-oriented, high-level programming language with dynamic semantics. Its high-level, built-in data structure, combined with dynamic typing and dynamic binding, makes it very attractive for Rapid Application Development, as well as for use as a scripting or glue language to connect existing components together. Python's simple, easy-to-learn syntax emphasizes readability and therefore reduces the cost of program maintenance.*"

If you are somewhat new to programming, the object-oriented, dynamic semantics mentioned previously probably do not mean much to you. But I think we can all agree that for rapid application development, simple, and easy-to-learn syntax sounds like a good thing. Python, as an interpreted language, means there is no compilation process required, so the time to write, test, and edit Python programs is greatly reduced. For simple scripts, if your script fails, a `print` statement is usually all you need to debug what was going on. Using the interpreter also means that Python is easily ported to different types of operating systems, such as Windows and Linux, and a Python program written on one operating system can be used on another.

The object-oriented nature encourages code reuse by breaking a large program into simple reusable objects, as well as other reusable formats with functions, modules, and packages. In fact, all Python files are modules that can be reused or imported into another Python program. This makes it easy to share programs between engineers and encourages code reuse. Python also has a *batteries included* mantra, which means that for common tasks, you need not download any additional packages. In order to achieve this without the code being too bloated, a set of standard libraries is installed when you install the Python interpreter. For common tasks such as regular expression, mathematics functions, and JSON decoding, all you need is the `import` statement, and the interpreter will move those functions into your program. This is what I would consider one of the killer features of the Python language.

Lastly, the fact that Python code can start in a relatively small-sized script with a few lines of code and grow into a full production system is very handy for network engineers. As many of us know, the network typically grows organically without a master plan. A language that can grow with your network in size is invaluable. You might be surprised to see a language that was deemed as a scripting language by many is being used for full production systems by many cutting-edge companies (organizations using Python, https://wiki.python.org/moin/OrganizationsUsingPython).

If you have ever worked in an environment where you have to switch between working on different vendor platforms, such as Cisco IOS and Juniper Junos, you know how painful it is to switch between syntaxes and usage when trying to achieve the same task. With Python being flexible enough for large and small programs, there is no such context switching, because it is just Python.

For the rest of the chapter, we will take a high-level tour of the Python language for a bit of a refresher. If you are already familiar with the basics, feel free to quickly scan through it or skip the rest of the chapter.

Python versions

As many readers are already aware, Python has been going through a transition from Python 2 to Python 3 for the last few years. Python 3 was released back in 2008, over 10 years ago, with active development with the most recent release of 3.7. Unfortunately, Python 3 is not backward compatible with Python 2. At the time of writing the second edition of this book, in the middle of 2018, the Python community has largely moved over to Python 3. The latest Python 2.x release, 2.7, was released over six years ago in mid-2010. Fortunately, both versions can coexist on the same machine. Personally, I use Python 2 as my default interpreter when I type in Python at the Command Prompt, and I use Python 3 when I need to use Python 3. More information is given in the next section about invoking Python interpreter, but here is an example of invoking Python 2 and Python 3 on an Ubuntu Linux machine:

```
$ python
Python 2.7.12 (default, Dec 4 2017, 14:50:18)
[GCC 5.4.0 20160609] on linux2
Type "help", "copyright", "credits" or "license" for more information.
>>> exit()

$ python3
Python 3.5.2 (default, Nov 23 2017, 16:37:01)
[GCC 5.4.0 20160609] on linux
Type "help", "copyright", "credits" or "license" for more information.
>>> exit()
```

With the 2.7 release being end-of-life, most Python frameworks now support Python 3. Python 3 also has lots of good features, such as asynchronous I/O that can be taken advantage of when we need to optimize our code. This book will use Python 3 for its code examples unless otherwise stated. We will also try to point out the Python 2 and Python 3 differences when applicable.

If a particular library or framework is better suited for Python 2, such as Ansible (see the following information), it will be pointed out and we will use Python 2 instead.

At the time of writing, Ansible 2.5 and above have support for Python 3. Prior to 2.5, Python 3 support was considered a tech preview. Given the relatively new supportability, many of the community modules are still yet to migrate to Python 3. For more information on Ansible and Python 3, please see https://docs.ansible.com/ansible/2.5/dev_guide/developing_python_3.html.

Operating system

As mentioned, Python is cross-platform. Python programs can be run on Windows, Mac, and Linux. In reality, certain care needs to be taken when you need to ensure cross-platform compatibility, such as taking care of the subtle difference between backslashes in Windows filenames. Since this book is for DevOps, systems, and network engineers, Linux is the preferred platform for the intended audience, especially in production. The code in this book will be tested on the Linux Ubuntu 16.06 LTS machine. I will also try my best to make sure the code runs the same on the Windows and the MacOS platform.

If you are interested in the OS details, they are as follows:

```
$ uname -a
Linux packt-network-python 4.13.0-45-generic #50~16.04.1-Ubuntu SMP Wed May 30 11:18:27 UTC 2018 x86_64 x86_64 x86_64 GNU/Linux
```

Running a Python program

Python programs are executed by an interpreter, which means the code is fed through this interpreter to be executed by the underlying operating system and results are displayed. There are several different implementations of the interpreter by the Python development community, such as IronPython and Jython. In this book, we will use the most common Python interpreter in use today, CPython. Whenever we mention Python in this book, we are referring to CPython unless otherwise indicated.

One way you can use Python is by taking advantage of the interactive prompt. This is useful when you want to quickly test a piece of Python code or concept without writing a whole program. This is typically done by simply typing in the `Python` keyword:

```
Python 3.5.2 (default, Nov 17 2016, 17:05:23)
[GCC 5.4.0 20160609] on linux
Type "help", "copyright", "credits" or "license" for more information.
>>> print("hello world")
hello world
>>>
```

In Python 3, the `print` statement is a function; therefore, it requires parentheses. In Python 2, you can omit the parentheses.

The interactive mode is one of Python's most useful features. In the interactive shell, you can type any valid statement or sequence of statements and immediately get a result back. I typically use this to explore a feature or library that I am not familiar with. Talk about instant gratification!

 On Windows, if you do not get a Python shell prompt back, you might not have the program in your system search path. The latest Windows Python installation program provides a checkbox for adding Python to your system path; make sure that is checked. Or you can add the program in the path manually by going to **Environment Settings**.

A more common way to run the Python program, however, is to save your Python file and run it via the interpreter after. This will save you from typing in the same statements over and over again like you have to do in the interactive shell. Python files are just regular text files that are typically saved with the .py extension. In the *Nix world, you can also add the **shebang** (#!) line on top to specify the interpreter that will be used to run the file. The # character can be used to specify comments that will not be executed by the interpreter. The following file, helloworld.py, has the following statements:

```
# This is a comment
print("hello world")
```

This can be executed as follows:

```
$ python helloworld.py
hello world
$
```

Python built-in types

Python has several standard types built in to the interpreter:

- **None**: The `Null` object
- **Numerics**: `int`, `long`, `float`, `complex`, and `bool` (the subclass of `int` with a `True` or `False` value)
- **Sequences**: `str`, list, tuple, and range
- **Mappings**: `dict`
- **Sets**: `set` and `frozenset`

The None type

The None type denotes an object with no value. The None type is returned in functions that do not explicitly return anything. The None type is also used in function arguments to error out if the caller does not pass in an actual value.

Numerics

Python numeric objects are basically numbers. With the exception of Booleans, the numeric types of int, long, float, and complex are all signed, meaning they can be positive or negative. A Boolean is a subclass of the integer, which can be one of two values: 1 for True, and 0 for False. The rest of the numeric types are differentiated by how precisely they can represent the number; for example, int are whole numbers with a limited range while long are whole numbers with unlimited range. Floats are numbers using the double-precision representation (64-bit) on the machine.

Sequences

Sequences are ordered sets of objects with an index of non-negative integers. In this and the next few sections, we will use the interactive interpreter to illustrate the different types. Please feel free to type along on your own computer.

Sometimes it surprises people that string is actually a sequence type. But if you look closely, strings are a series of characters put together. Strings are enclosed by either single, double, or triple quotes. Note in the following examples, the quotes have to match, and triple quotes allow the string to span different lines:

```
>>> a = "networking is fun"
>>> b = 'DevOps is fun too'
>>> c = """what about coding?
... super fun!"""
>>>
```

The other two commonly used sequence types are lists and tuples. Lists are sequences of arbitrary objects. Lists can be created by enclosing objects in square brackets. Just like strings, lists are indexed by non-zero integers that start at zero. The values of lists are retrieved by referencing the index number:

```
>>> vendors = ["Cisco", "Arista", "Juniper"]
>>> vendors[0]
'Cisco'
>>> vendors[1]
```

```
'Arista'
>>> vendors[2]
'Juniper'
```

Tuples are similar to lists, created by enclosing values in parentheses. Like lists, the values in the tuple are retrieved by referencing its index number. Unlike lists, the values cannot be modified after creation:

```
>>> datacenters = ("SJC1", "LAX1", "SFO1")
>>> datacenters[0]
'SJC1'
>>> datacenters[1]
'LAX1'
>>> datacenters[2]
'SFO1'
```

Some operations are common to all sequence types, such as returning an element by index as well as slicing:

```
>>> a
'networking is fun'
>>> a[1]
'e'
>>> vendors
['Cisco', 'Arista', 'Juniper']
>>> vendors[1]
'Arista'
>>> datacenters
('SJC1', 'LAX1', 'SFO1')
>>> datacenters[1]
'LAX1'
>>>
>>> a[0:2]
'ne'
>>> vendors[0:2]
['Cisco', 'Arista']
>>> datacenters[0:2]
('SJC1', 'LAX1')
>>>
```

Remember that index starts at 0. Therefore, the index of 1 is actually the second element in the sequence.

Chapter 1

There are also common functions that can be applied to sequence types, such as checking the number of elements and the minimum and maximum values:

```
>>> len(a)
17
>>> len(vendors)
3
>>> len(datacenters)
3
>>>
>>> b = [1, 2, 3, 4, 5]
>>> min(b)
1
>>> max(b)
5
```

It will come as no surprise that there are various methods that apply only to strings. It is worth noting that these methods do not modify the underlying string data itself and always return a new string. If you want to use the new value, you will need to catch the return value and assign it to a different variable:

```
>>> a
'networking is fun'
>>> a.capitalize()
'Networking is fun'
>>> a.upper()
'NETWORKING IS FUN'
>>> a
'networking is fun'
>>> b = a.upper()
>>> b
'NETWORKING IS FUN'
>>> a.split()
['networking', 'is', 'fun']
>>> a
'networking is fun'
>>> b = a.split()
>>> b
['networking', 'is', 'fun']
>>>
```

Here are some of the common methods for a list. This list is a very useful structure in terms of putting multiple items together and iterating through them one at a time. For example, we can make a list of data center spine switches and apply the same access list to all of them by iterating through them one by one. Since a list's value can be modified after creation (unlike tuples), we can also expand and contrast the existing list as we move along the program:

```
>>> routers = ['r1', 'r2', 'r3', 'r4', 'r5']
>>> routers.append('r6')
>>> routers
['r1', 'r2', 'r3', 'r4', 'r5', 'r6']
>>> routers.insert(2, 'r100')
>>> routers
['r1', 'r2', 'r100', 'r3', 'r4', 'r5', 'r6']
>>> routers.pop(1)
'r2'
>>> routers
['r1', 'r100', 'r3', 'r4', 'r5', 'r6']
```

Mapping

Python provides one mapping type, called the **dictionary**. The dictionary is what I think of as a poor man's database because it contains objects that can be indexed by keys. This is often referred to as the associated array or hashing table in other languages. If you have used any of the dictionary-like objects in other languages, you will know that this is a powerful type, because you can refer to the object with a human-readable key. This key will make more sense for the poor guy who is trying to maintain and troubleshoot the code. That guy could be you only a few months after you wrote the code and troubleshooting at 2 a.m.. The object in the dictionary value can also be another data type, such as a list. You can create a dictionary with curly braces:

```
>>> datacenter1 = {'spines': ['r1', 'r2', 'r3', 'r4']}
>>> datacenter1['leafs'] = ['l1', 'l2', 'l3', 'l4']
>>> datacenter1
{'leafs': ['l1', 'l2', 'l3', 'l4'], 'spines': ['r1', 'r2', 'r3', 'r4']}
>>> datacenter1['spines']
['r1', 'r2', 'r3', 'r4']
>>> datacenter1['leafs']
['l1', 'l2', 'l3', 'l4']
```

Sets

A **set** is used to contain an unordered collection of objects. Unlike lists and tuples, sets are unordered and cannot be indexed by numbers. But there is one character that makes sets standout as useful: the elements of a set are never duplicated. Imagine you have a list of IPs that you need to put in an access list of. The only problem in this list of IPs is that they are full of duplicates. Now, think about how many lines of code you would use to loop through the list of IPs to sort out unique items, one at a time. However, the built-in set type would allow you to eliminate the duplicate entries with just one line of code. To be honest, I do not use set that much, but when I need it, I am always very thankful it exists. Once the set or sets are created, they can be compared with each other using the union, intersection, and differences:

```
>>> a = "hello"
>>> set(a)
{'h', 'l', 'o', 'e'}
>>> b = set([1, 1, 2, 2, 3, 3, 4, 4])
>>> b
{1, 2, 3, 4}
>>> b.add(5)
>>> b
{1, 2, 3, 4, 5}
>>> b.update(['a', 'a', 'b', 'b'])
>>> b
{1, 2, 3, 4, 5, 'b', 'a'}
>>> a = set([1, 2, 3, 4, 5])
>>> b = set([4, 5, 6, 7, 8])
>>> a.intersection(b)
{4, 5}
>>> a.union(b)
{1, 2, 3, 4, 5, 6, 7, 8}
>>> 1 *
{1, 2, 3}
>>>
```

Python operators

Python has some numeric operators that you would expect; note that the truncating division, (//, also known as **floor division**) truncates the result to an integer and a floating point and returns the integer value. The modulo (%) operator returns the remainder value in the division:

```
>>> 1 + 2
3
```

```
>>> 2 - 1
1
>>> 1 * 5
5
>>> 5 / 1
5.0
>>> 5 // 2
2
>>> 5 % 2
1
```

There are also comparison operators. Note the double equals sign for comparison and a single equals sign for variable assignment:

```
>>> a = 1
>>> b = 2
>>> a == b
False
>>> a > b
False
>>> a < b
True
>>> a <= b
True
```

We can also use two of the common membership operators to see whether an object is in a sequence type:

```
>>> a = 'hello world'
>>> 'h' in a
True
>>> 'z' in a
False
>>> 'h' not in a
False
>>> 'z' not in a
True
```

Python control flow tools

The `if`, `else`, and `elif` statements control conditional code execution. As one would expect, the format of the conditional statement is as follows:

```
if expression:
    do something
elif expression:
```

```
    do something if the expression meets
elif expression:
    do something if the expression meets
...
else:
    statement
```

Here is a simple example:

```
>>> a = 10
>>> if a > 1:
...     print("a is larger than 1")
... elif a < 1:
...     print("a is smaller than 1")
... else:
...     print("a is equal to 1")
...
a is larger than 1
>>>
```

The `while` loop will continue to execute until the condition is `false`, so be careful with this one if you don't want to continue to execute (and crash your process):

```
while expression:
    do something
>>> a = 10
>>> b = 1
>>> while b < a:
...     print(b)
...     b += 1
...
1
2
3
4
5
6
7
8
9
```

The `for` loop works with any object that supports iteration; this means all the built-in sequence types, such as lists, tuples, and strings, can be used in a `for` loop. The letter `i` in the following `for` loop is an iterating variable, so you can typically pick something that makes sense within the context of your code:

```
for i in sequence:
  do something

>>> a = [100, 200, 300, 400]
>>> for number in a:
...     print(number)
...
100
200
300
400
```

You can also make your own object that supports the iterator protocol and be able to use the `for` loop for this object.

Constructing such an object is outside the scope of this chapter, but it is useful knowledge to have; you can read more about it at `https://docs.python.org/3/c-api/iter.html`.

Python functions

Most of the time, when you find yourself copy and pasting some pieces of code, you should break it up into a self-contained chunk into functions. This practice allows for better modularity, is easier to maintain, and allows for code reuse. Python functions are defined using the `def` keyword with the function name, followed by the function parameters. The body of the function consists of the Python statements that are to be executed. At the end of the function, you can choose to return a value to the function caller, or by default, it will return the `None` object if you do not specify a return value:

```
def name(parameter1, parameter2):
    statements
    return value
```

We will see a lot more examples of function in the following chapters, so here is a quick example:

```
>>> def subtract(a, b):
...     c = a - b
```

```
...         return c
...
>>> result = subtract(10, 5)
>>> result
5
>>>
```

Python classes

Python is an **object-oriented programming** (**OOP**) language. The way Python creates objects is with the `class` keyword. A Python object is most commonly a collection of functions (methods), variables, and attributes (properties). Once a class is defined, you can create instances of such a class. The class serves as a blueprint for subsequent instances.

The topic of OOP is outside the scope of this chapter, so here is a simple example of a `router` object definition:

```
>>> class router(object):
...     def __init__(self, name, interface_number, vendor):
...         self.name = name
...         self.interface_number = interface_number
...         self.vendor = vendor
...
>>>
```

Once defined, you are able to create as many instances of that class as you'd like:

```
>>> r1 = router("SFO1-R1", 64, "Cisco")
>>> r1.name
'SFO1-R1'
>>> r1.interface_number
64
>>> r1.vendor
'Cisco'
>>>
>>> r2 = router("LAX-R2", 32, "Juniper")
>>> r2.name
'LAX-R2'
>>> r2.interface_number
32
>>> r2.vendor
'Juniper'
>>>
```

Of course, there is a lot more to Python objects and OOP. We will look at more examples in future chapters.

Python modules and packages

Any Python source file can be used as a module, and any functions and classes you define in that source file can be reused. To load the code, the file referencing the module needs to use the `import` keyword. Three things happen when the file is imported:

1. The file creates a new namespace for the objects defined in the source file
2. The caller executes all the code contained in the module
3. The file creates a name within the caller that refers to the module being imported. The name matches the name of the module

Remember the `subtract()` function that you defined using the interactive shell? To reuse the function, we can put it into a file named `subtract.py`:

```
def subtract(a, b):
    c = a - b
    return c
```

In a file within the same directory of `subtract.py`, you can start the Python interpreter and import this function:

```
Python 2.7.12 (default, Nov 19 2016, 06:48:10)
[GCC 5.4.0 20160609] on linux2
Type "help", "copyright", "credits" or "license" for more information.
>>> import subtract
>>> result = subtract.subtract(10, 5)
>>> result
5
```

This works because, by default, Python will first search for the current directory for the available modules. If you are in a different directory, you can manually add a search path location using the `sys` module with `sys.path`. Remember the standard library that we mentioned a while back? You guessed it, those are just Python files being used as modules.

Packages allow a collection of modules to be grouped together. This further organizes Python modules into a more namespace protection to further reusability. A package is defined by creating a directory with a name you want to use as the namespace, then you can place the module source file under that directory. In order for Python to recognize it as a Python-package, just create a __init__.py file in this directory. In the same example as the subtract.py file, if you were to create a directory called math_stuff and create a __init__.py file:

```
echou@pythonicNeteng:~/Master_Python_Networking/Chapter1$ mkdir math_stuff
echou@pythonicNeteng:~/Master_Python_Networking/Chapter1$ touch math_stuff/__init__.py
echou@pythonicNeteng:~/Master_Python_Networking/Chapter1$ tree .
.
├── helloworld.py
└── math_stuff
    ├── __init__.py
    └── subtract.py

1 directory, 3 files
echou@pythonicNeteng:~/Master_Python_Networking/Chapter1$
```

The way you will now refer to the module will need to include the package name:

```
>>> from math_stuff.subtract import subtract
>>> result = subtract(10, 5)
>>> result
5
>>>
```

As you can see, modules and packages are great ways to organize large code files and make sharing Python code a lot easier.

Summary

In this chapter, we covered the OSI model and reviewed network protocol suites, such as TCP, UDP, and IP. They work as the layers that handle the addressing and communication negotiation between any two hosts. The protocols were designed with extensibility in mind and have largely been unchanged from their original design. Considering the explosive growth of the internet, that is quite an accomplishment.

We also quickly reviewed the Python language, including built-in types, operators, control flows, functions, classes, modules, and packages. Python is a powerful, production-ready language that is also easy to read. This makes the language an ideal choice when it comes to network automation. Network engineers can leverage Python to start with simple scripts and gradually move on to other advanced features.

In `Chapter 2`, *Low-Level Network Device Interactions*, we will start to look at using Python to programmatically interact with network equipment.

Low-Level Network Device Interactions

In Chapter 1, *Review of TCP/IP Protocol Suite and Python*, we looked at the theories and specifications behind network communication protocols. We also took a quick tour of the Python language. In this chapter, we will start to dive deeper into the management of network devices using Python. In particular, we will examine the different ways in which we can use Python to programmatically communicate with legacy network routers and switches.

What do I mean by legacy network routers and switches? While it is hard to imagine any networking device coming out today without an **Application Program Interface** (**API**) for programmatic communication, it is a known fact that many of the network devices deployed in previous years did not contain API interfaces. The intended method of management for those devices was through **Command Line Interfaces** (**CLIs**) using terminal programs, which were originally developed with a human engineer in mind. The management relied on the engineer's interpretation of the data returned from the device for appropriate action. As the number of network devices and the complexity of the network grew, it became increasingly difficult to manually manage them one by one.

Python has two great libraries that can help with these tasks, Pexpect and Paramiko, as well as other libraries derived from them. This chapter will cover Pexpect first, then move on with examples from Paramiko. Once you understand the basics of Paramiko, it is easy to branch out to expanded libraries such as Netmiko. It is also worth mentioning that Ansible (covered in Chapters 4, *The Python Automation Framework – Ansible Basics*, and Chapter 5, *The Python Automation Framework – Beyond Basics*) relies heavily on Paramiko for its network modules. In this chapter, we will take a look at the following topics:

- The challenges of the CLI
- Constructing a virtual lab
- The Python Pexpect library

- The Python Paramiko library
- The downsides of Pexpect and Paramiko

Let's get started!

The challenges of the CLI

At the Interop expo in Las Vegas in 2014, *BigSwitch Networks'* CEO Douglas Murray displayed the following slide to illustrate what had changed in **Data Center Networking** (**DCN**) in the 20 years between 1993 to 2013:

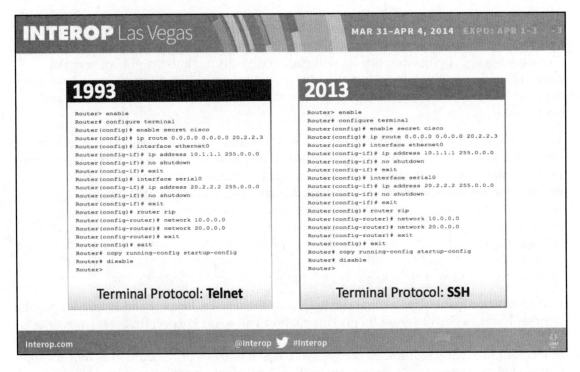

Data center networking changes (source: https://www.bigswitch.com/sites/default/files/presentations/murraydouglasstartuphotseatpanel.pdf)

His point was apparent: not much had changed in those 20 years in the way we manage network devices. While he might have been negatively biased toward the incumbent vendors when displaying this slide, his point is well taken. In his opinion, the only thing that had changed about managing routers and switches in 20 years was the protocol changing from the less secure Telnet to the more secure SSH.

It was right around the same time in 2014 that we started to see the industry coming to a consensus about the clear need to move away from manual, human-driven CLI toward an automatic, computer-centric automation API. Make no mistake, we still need to directly communicate with the device when making network designs, bringing up initial proof of concepts, and deploying the topology for the first time. However, once we have moved beyond the initial deployment, the requirement is to consistently make the same changes reliably, to make them error-free, and to repeat them over and over again without the engineer being distracted or feeling tired. This requirement sounds like an ideal job for computers and our favorite language, Python.

Referring back to the slide, the main challenge is the interaction between the router and the administrator. The router will output a series of information and will expect the administrator to enter a series of manual commands from the engineer's interpretation of the output. For example, you have to type in `enable` to get into a privileged mode, and upon receiving the returned prompt with the # sign, you then type in `configure terminal` in order to go into the configuration mode. The same process can further be expanded into the interface configuration mode and routing protocol configuration mode. This is in sharp contrast to a computer-driven, programmatic mindset. When the computer wants to accomplish a single task, say, put an IP address on an interface, it wants to structurally give all the information to the router at once, and it would expect a single `yes` or `no` answer from the router to indicate the success or failure of the task.

The solution, as implemented by both Pexpect and Paramiko, is to treat the interactive process as a child process and watch over the interaction between the process and the destination device. Based on the returned value, the parent process will decide the subsequent action, if any.

Constructing a virtual lab

Before we dive into the packages, let's examine the options of putting together a lab for the benefit of learning. As the old saying goes, *practice makes perfect*: we need an isolated sandbox to safely make mistakes, try out new ways of doing things, and repeat some of the steps to reinforce concepts that were not clear in the first try. It is easy enough to install Python and the necessary packages for the management host, but what about those routers and switches that we want to simulate?

To put together a network lab, we basically have two options, each with its advantages and disadvantages:

- **Physical device**: This option consists of physical devices that you can see and touch. If you are lucky enough, you might be able to put together a lab that is an exact replication of your production environment:
    - **Advantages**: It is an easy transition from lab to production, easier to understand by managers and fellow engineers who can look at and touch the devices. In short, the comfort level with physical devices is extremely high because of familiarity.
    - **Disadvantages**: It is relatively expensive to pay for a device that is only used in the lab. Devices require engineering hours to rack and stack and are not very flexible once constructed.
- **Virtual devices**: These are emulations or simulations of actual network devices. They are either provided by the vendors or by the open source community:
    - **Advantages**: Virtual devices are easier to set up, relatively cheap, and can make changes to the topology quickly.
    - **Disadvantages**: They are usually a scaled-down version of their physical counterpart. Sometimes there are feature gaps between the virtual and the physical device.

Of course, deciding on a virtual or physical lab is a personal decision derived from a trade-off between the cost, ease of implementation, and the risk of having a gap between lab and production. In some of the environments I have worked on, the virtual lab is used when doing an initial proof-of-concept while the physical lab is used when we move closer to the final design.

In my opinion, as more and more vendors decide to produce virtual appliances, the virtual lab is the way to proceed in a learning environment. The feature gap of the virtual appliance is relatively small and specifically documented, especially when the virtual instance is provided by the vendor. The cost of the virtual appliance is relatively small compared to buying physical devices. The time-to-build using virtual devices is quicker because they are usually just software programs.

For this book, I will use a combination of physical and virtual devices for concept demonstration with a preference for virtual devices. For the examples we will see, the differences should be transparent. If there are any known differences between the virtual and physical devices pertaining to our objectives, I will make sure to list them.

On the virtual lab front, besides images from various vendors, I am using a program from Cisco called **Virtual Internet Routing Lab (VIRL)**, `https://learningnetworkstore.cisco.com/virtual-internet-routing-lab-virl/cisco-personal-edition-pe-20-nodes-virl-20`.

I want to point out that the use of this program is entirely optional for the reader. But it is strongly recommended that the reader have some lab equipment to follow along with the examples in this book.

Cisco VIRL

I remember when I first started to study for my **Cisco Certified Internetwork Expert** (**CCIE**) lab exam, I purchased some used Cisco equipment from eBay to study with. Even at a discount, each router and switch cost hundreds of US dollars, so to save money, I purchased some really outdated Cisco routers from the 1980s (search for Cisco AGS routers in your favorite search engine for a good chuckle), which significantly lacked features and horsepower, even for lab standards. As much as it made for an interesting conversation with family members when I turned them on (they were really loud), putting the physical devices together was not fun. They were heavy and clunky, it was a pain to connect all the cables, and to introduce link failure, I would literally unplug a cable.

Fast-forward a few years. **Dynamip** was created and I fell in love with how easy it was to create different network scenarios. This was especially important when I tried to learn a new concept. All you need is the IOS images from Cisco, a few carefully constructed topology files, and you can easily construct a virtual network that you can test your knowledge on. I had a whole folder of network topologies, pre-saved configurations, and different version of images, as called for by the scenario. The addition of a GNS3 frontend gives the whole setup a beautiful GUI facelift. With GNS3, you can just click and drop your links and devices; you can even just print out the network topology for your manager right out of the GNS3 design panel. The only thing that was lacking was the tool not being officially blessed by the vendor and the perceived lack of credibility because of it.

In 2015, the Cisco community decided to fulfill this need by releasing the Cisco VIRL. If you have a server that meets the requirements and you are willing to pay for the required annual license, this is my preferred method of developing and trying out much of the Python code, both for this book and my own production use.

Low-Level Network Device Interactions

As of January 1 2017, only the personal edition 20-Node license is available for purchase for USD $199.99 per year.

Even at a monetary cost, in my opinion, the VIRL platform offers a few advantages over other alternatives:

- **Ease of use**: All the images for IOSv, IOS-XRv, CSR100v, NX-OSv, and ASAv are included in a single download.
- **Official (kind of)**: Although support is community-driven, it is a widely used tool internally at Cisco. Because of its popularity, the bugs get fixed quickly, new features are carefully documented, and useful knowledge is widely shared among its users.
- **The cloud migration path**: The project offers a logical migration path when your emulation grows out of the hardware power you have, such as Cisco dCloud (https://dcloud.cisco.com/), VIRL on Packet (http://virl.cisco.com/cloud/), and Cisco DevNet (https://developer.cisco.com/). This is an important feature that sometimes gets overlooked.
- **The link and control-plane simulation**: The tool can simulate latency, jitter, and packet loss on a per-link basis for real-world link characteristics. There is also a control-plane traffic generator for external route injection.
- **Others**: The tool offers some nice features, such as VM Maestro topology design and simulation control, AutoNetkit for automatic config generation, and user workspace management if the server is shared. There are also open source projects such as virlutils (https://github.com/CiscoDevNet/virlutils), which are actively worked on by the community to enhance the workability of the tool.

We will not use all of the features in VIRL in this book. But since this is a relatively new tool that is worth your consideration, if you do decide this is the tool you would like to use, I want to offer some of the setups I used.

Again, I want to stress the importance of having a lab, but it does not need to be the Cisco VIRL lab. The code examples provided in this book should work across any lab device, as long as they run the same software type and version.

VIRL tips

The VIRL website (http://virl.cisco.com/) offers lots of guidance, preparation, and documentation. I also find that the VIRL user community generally offers quick and accurate help. I will not repeat information already offered in those two places; however, here are some of the setups I use for the lab in this book:

1. VIRL uses two virtual Ethernet interfaces for connections. The first interface is set up as NAT for the host machine's internet connection, and the second is used for local management interface connectivity (VMnet2 in the following example). I use a separate virtual machine with a similar network setup in order to run my Python code, with the first primary Ethernet used for internet connectivity and the second Ethernet connection to Vmnet2 for lab device management network:

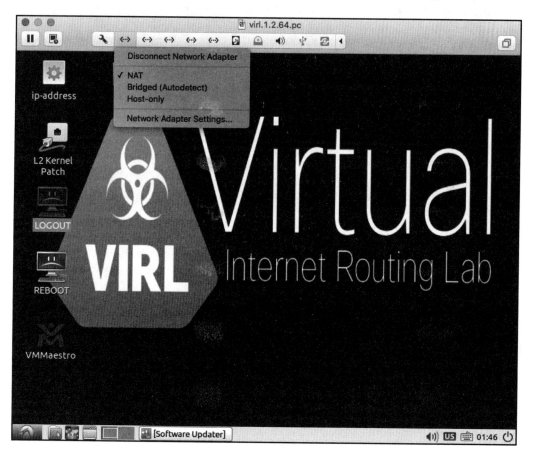

Low-Level Network Device Interactions

2. VMnet2 is a custom network created to connect the Ubuntu host with the VIRL virtual machine:

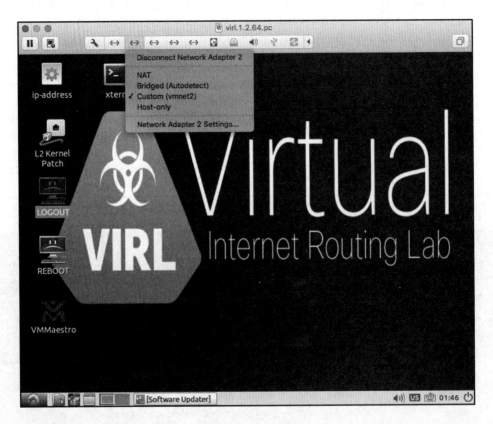

3. In the **Topology Design** option, I set the **Management Network** option to **Shared flat network** in order to use VMnet2 as the management network on the virtual routers:

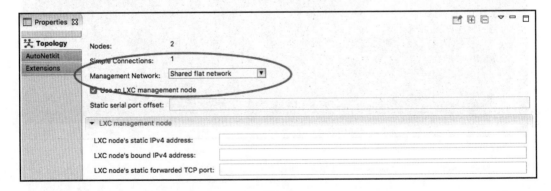

Chapter 2

4. Under the node configuration, you have the option to statically configure the management IP. I try to statically set the management IP addresses instead of having them dynamically assigned by the software. This allows for more deterministic accessibility:

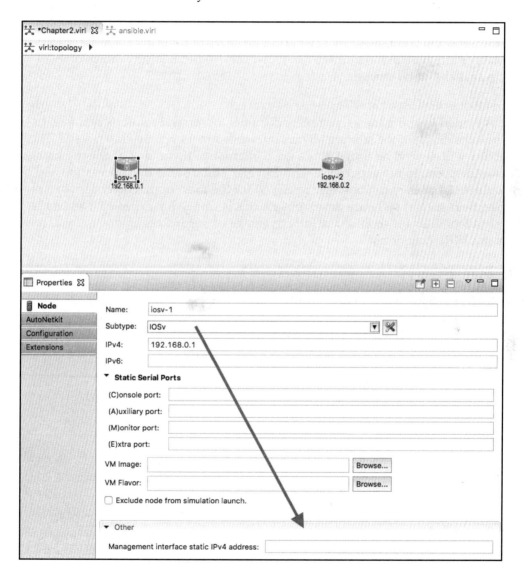

Cisco DevNet and dCloud

Cisco provides two other excellent, and, at the time of writing, free, methods for practicing network automation with various Cisco gears. Both of the tools require a **Cisco Connection Online (CCO)** login. They are both really good, especially for the price point (they are free!). It is hard for me to imagine that these online tools will remain free for long; it is my belief that, at some point, these tools will need to charge money for their usage or be rolled into a bigger initiative that requires a fee. However, we can take advantage of them while they are available at no charge.

The first tool is the Cisco DevNet (`https://developer.cisco.com/`) sandbox, which includes guided learning tracks, complete documentation, and sandbox remote labs, among other benefits. Some of the labs are always on, while others you need to reserve. The lab availability will depend on usage. It is a great alternative if you do not already have a lab at your own disposal. In my experience with DevNet, some of the documentation and links were outdated, but they can be easily retrieved for the most updated version. In a rapidly changing field such as software development, this is somewhat expected. DevNet is certainly a tool that you should take full advantage of, regardless of whether you have a locally run VIRL host or not:

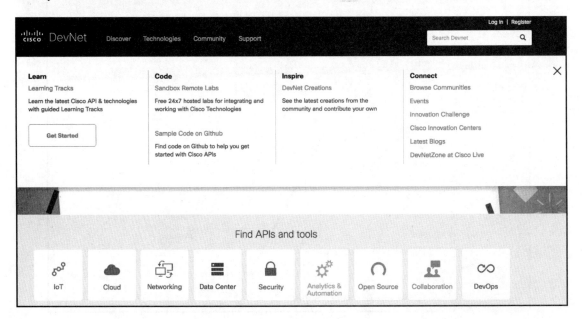

Another online lab option for Cisco is `https://dcloud.cisco.com/`. You can think of dCloud as running VIRL on other people's servers without having to manage or pay for those resources. It seems that Cisco is treating dCloud as both a standalone product as well as an extension to VIRL. For example, in the use case of when you are unable to run more than a few IOX-XR or NX-OS instances locally, you can use dCloud to extend your local lab. It is a relatively new tool, but it is definitely worth a look:

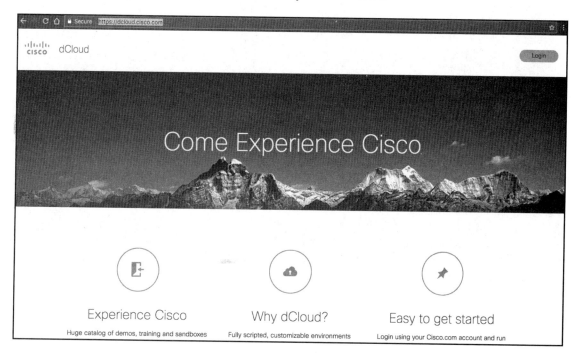

GNS3

There are a few other virtual labs that I use for this book and other purposes. The `GNS3` tool is one of them:

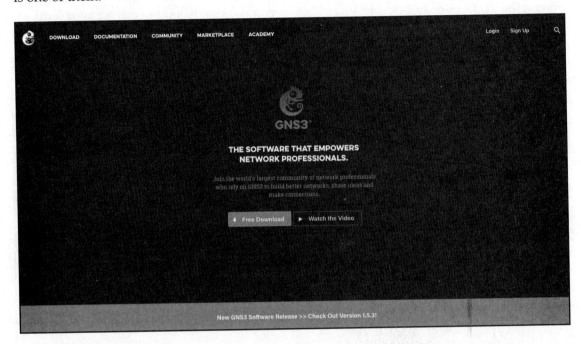

As mentioned previously in this chapter, **GNS3** is what a lot of us used to study for certification tests and to practice for labs. The tool has really grown up from the early days of the simple frontend for Dynamips into a viable commercial product. Cisco-made tools, such as VIRL, DevNet, and dCloud, only contain Cisco technologies. Even though they provide ways for virtual lab devices to communicate with the outside world, they are not as easy as just having multi-vendor virtualized appliances living directly in the simulation environment. **GNS3** is vendor-neutral and can include a multi-vendor virtualized platform directly in the lab. This is typically done either by making a clone of the image (such as Arista vEOS) or by directly launching the network device image via other hypervisors (such as Juniper Olive emulation). Some might argue that GNS3 does not have the breadth and depth of the Cisco VIRL project, but since they can run different variation Cisco technologies, I often use it when I need to incorporate other vendor technologies into the lab.

Another multi-vendor network emulation environment that has gotten a lot of great reviews is the **Emulated Virtual Environment Next Generation (EVE-NG)**, http://www.eve-ng.net/. I personally do not have much experience with the tool, but many of my colleagues and friends in the industry use it for their network labs.

There are also other virtualized platforms, such as Arista vEOS (https://eos.arista.com/tag/veos/), Juniper vMX (http://www.juniper.net/us/en/products-services/routing/mx-series/vmx/), and vSRX (http://www.juniper.net/us/en/products-services/security/srx-series/vsrx/), which you can use as a standalone virtual appliance during testing. They are great complementary tools for testing platform-specific features, such as the differences between the API versions on the platform. Many of them are offered as paid products on public cloud provider marketplaces for easier access. They are often offered the identical feature as their physical counterpart.

Python Pexpect library

> *Pexpect is a pure Python module for spawning child applications, controlling them, and responding to expected patterns in their output. Pexpect works like Don Libes' Expect. Pexpct allows your script to spawn a child application and control it as if a human were typing commands. Pexpect, Read the Docs:* https://pexpect.readthedocs.io/en/stable/index.html

Let's take a look at the Python Pexpect library. Similar to the original Tcl Expect module by Don Libe, Pexpect launches or spawns another process and watches over it in order to control the interaction. The Expect tool was originally developed to automate interactive processes such as FTP, Telnet, and rlogin, and was later expanded to include network automation. Unlike the original Expect, Pexpect is entirely written in Python, which does not require TCL or C extensions to be compiled. This allows us to use the familiar Python syntax and its rich standard library in our code.

Pexpect installation

Since this is the first package we will install, we will install both the `pip` tool with the `pexpect` package. The process is pretty straightforward:

```
sudo apt-get install python-pip #Python2
sudo apt-get install python3-pip
sudo pip3 install pexpect
sudo pip install pexpect #Python2
```

Low-Level Network Device Interactions

 I am using `pip3` to install Python 3 packages, while using `pip` to install packages in the Python 2 environment.

Do a quick to test to make sure the package is usable:

```
>>> import pexpect
>>> dir(pexpect)
['EOF', 'ExceptionPexpect', 'Expecter', 'PY3',
 'TIMEOUT', '__all__', '__builtins__', '__cached__',
 '__doc__', '__file__', '__loader__', '__name__',
 '__package__', '__path__', '__revision__',
 '__spec__', '__version__', 'exceptions', 'expect',
 'is_executable_file', 'pty_spawn', 'run', 'runu',
 'searcher_re', 'searcher_string', 'spawn',
 'spawnbase', 'spawnu', 'split_command_line', 'sys',
 'utils', 'which']
>>>
```

Pexpect overview

For our first lab, we will construct a simple network with two IOSv devices connected back to back:

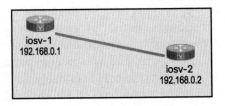

Lab topology

The devices will each have a loopback address in the `192.16.0.x/24` range and the management IP will be in the `172.16.1.x/24` range. The VIRL topology file is included in the accommodated book downloadable files. You can import the topology to your own VIRL software. If you do not have VIRL, you can also view the necessary information by opening the topology file with a text editor. The file is simply an XML file with each node's information under the `node` element:

Lab node information

With the devices ready, let's take a look at how you would interact with the router if you were to Telnet into the device:

```
echou@ubuntu:~$ telnet 172.16.1.20
Trying 172.16.1.20...
Connected to 172.16.1.20.
Escape character is '^]'.
<skip>
User Access Verification

Username: cisco
Password:
```

I used VIRL AutoNetkit to automatically generate the initial configuration of the routers, which generated the default username cisco, and the password cisco. Notice that the user is already in privileged mode because of the privilege assigned in the configuration:

```
iosv-1#sh run | i cisco
enable password cisco
username cisco privilege 15 secret 5 $1$Wiwq$7xt2oE0P9ThdxFS02trFw.
```

Low-Level Network Device Interactions

```
 password cisco
 password cisco
iosv-1#
```

The auto-config also generated `vty` access for both Telnet and SSH:

```
line vty 0 4
 exec-timeout 720 0
 password cisco
 login local
 transport input telnet ssh
```

Let's see a Pexpect example using the Python interactive shell:

```
Python 3.5.2 (default, Nov 17 2016, 17:05:23)
[GCC 5.4.0 20160609] on linux
Type "help", "copyright", "credits" or "license" for more information.
>>> import pexpect
>>> child = pexpect.spawn('telnet 172.16.1.20')
>>> child.expect('Username')
0
>>> child.sendline('cisco')
6
>>> child.expect('Password')
0
>>> child.sendline('cisco')
6
>>> child.expect('iosv-1#')
0
>>> child.sendline('show version | i V')
19
>>> child.expect('iosv-1#')
0
>>> child.before
b'show version | i VrnCisco IOS Software, IOSv Software (VIOS-ADVENTERPRISEK9-M), Version 15.6(2)T, RELEASE SOFTWARE (fc2)rnProcessor board ID 9MM4BI7B0DSWK4OKV1IIRrn'
>>> child.sendline('exit')
5
>>> exit()
```

 Starting from Pexpect version 4.0, you can run Pexpect on a Windows platform. But, as noted in the Pexpect documentation, running Pexpect on Windows should be considered experimental for now.

Chapter 2

In the previous interactive example, Pexpect spawns off a child process and watches over it in an interactive fashion. There are two important methods shown in the example, `expect()` and `sendline()`. The `expect()` line indicates that the string the Pexpect process looks for as an indicator for when the returned string is considered done. This is the expected pattern. In our example, we knew the router had sent us all the information when the hostname prompt (`iosv-1#`) was returned. The `sendline()` method indicates which words should be sent to the remote device as the command. There is also a method called `send()` but `sendline()` includes a linefeed, which is similar to pressing the *Enter* key at the end of the words you sent in your previous telnet session. From the router's perspective, it is just as if someone typed in the text from a Terminal. In other words, we are tricking the routers into thinking they are interfacing with a human being when they are actually communicating with a computer.

The `before` and `after` properties will be set to the text printed by the child application. The `before` properties will be set to the text printed by the child application up to the expected pattern. The `after` string will contain the text that was matched by the expected pattern. In our case, the `before` text will be set to the output between the two expected matches (`iosv-1#`), including the `show version` command. The `after` text is the router hostname prompt:

```
>>> child.sendline('show version | i V')
19
>>> child.expect('iosv-1#')
0
>>> child.before
b'show version | i VrnCisco IOS Software, IOSv Software (VIOS-
ADVENTERPRISEK9-M), Version 15.6(2)T, RELEASE SOFTWARE (fc2)rnProcessor
board ID 9MM4BI7B0DSWK40KV1IIRrn'
>>> child.after
b'iosv-1#'
```

What would happen if you expected the wrong term? For example, if you typed in `username` instead of `Username` after spawning the child application, then the Pexpect process would look for a string of `username` from the child process. In that case, the Pexpect process would just hang because the word `username` would never be returned by the router. The session would eventually timeout, or you could manually exit out via *Ctrl + C*.

The `expect()` method waits for the child application to return a given string, so in the previous example, if you wanted to accommodate both lowercase and uppercase u, you could use the following term:

```
>>> child.expect('[Uu]sername')
```

Low-Level Network Device Interactions

The square bracket serves as an or operation that tells the child application to expect a lowercase or uppercase u followed by sername as the string. What we are telling the process is that we will accept either Username or username as the expected string.

 For more information on Python regular expressions, go to https://docs.python.org/3.5/library/re.html.

The expect() method can also contain a list of options instead of just a single string; these options can also be regular expression themselves. Going back to the previous example, you can use the following list of options to accommodate the two different possible strings:

```
>>> child.expect(['Username', 'username'])
```

Generally speaking, use the regular expression for a single expect string when you can fit the different hostname in a regular expression, whereas use the possible options if you need to catch completely different responses from the router, such as a password rejection. For example, if you use several different passwords for your login, you want to catch % Login invalid as well as the device prompt.

One important difference between Pexpect regular expressions and Python regular expressions is that the Pexpect matching is non-greedy, which means they will match as little as possible when using special characters. Because Pexpect performs regular expression on a stream, you cannot look ahead, as the child process generating the stream may not be finished. This means the special dollar sign character $ typically matching the end of the line is useless because .+ will always return no characters, and the .* pattern will match as little as possible. In general, just keep this in mind and be as specific as you can be on the expect match strings.

Let's consider the following scenario:

```
>>> child.sendline('show run | i hostname')
22
>>> child.expect('iosv-1')
0
>>> child.before
b'show run | i hostnamernhostname '
>>>
```

Hmm... Something is not quite right here. Compare it to the Terminal output before; the output you expect would be hostname iosv-1:

```
iosv-1#show run | i hostname
hostname iosv-1
```

[54]

```
iosv-1#
```

Taking a closer look at the expected string will reveal the mistake. In this case, we were missing the hash (#) sign behind the `iosv-1` hostname. Therefore, the child application treated the second part of the return string as the expected string:

```
>>> child.sendline('show run | i hostname')
22
>>> child.expect('iosv-1#')
0
>>> child.before
b'show run | i hostnamernhostname iosv-1rn'
>>>
```

You can see a pattern emerging from the usage of Pexpect after a few examples. The user maps out the sequence of interactions between the Pexpect process and the child application. With some Python variables and loops, we can start to construct a useful program that will help us gather information and make changes to network devices.

Our first Pexpect program

Our first program, `chapter2_1.py`, extends what we did in the last section with some additional code:

```
#!/usr/bin/python3

import pexpect

devices = {'iosv-1': {'prompt': 'iosv-1#', 'ip': '172.16.1.20'},
'iosv-2': {'prompt': 'iosv-2#', 'ip': '172.16.1.21'}}
username = 'cisco'
password = 'cisco'

for device in devices.keys():
    device_prompt = devices[device]['prompt']
    child = pexpect.spawn('telnet ' + devices[device]['ip'])
    child.expect('Username:')
    child.sendline(username)
    child.expect('Password:')
    child.sendline(password)
    child.expect(device_prompt)
    child.sendline('show version | i V')
    child.expect(device_prompt)
    print(child.before)
    child.sendline('exit')
```

Low-Level Network Device Interactions

We use a nested dictionary in line 5:

```
devices = {'iosv-1': {'prompt': 'iosv-1#', 'ip':
'172.16.1.20'}, 'iosv-2': {'prompt': 'iosv-2#',
'ip': '172.16.1.21'}}
```

The nested dictionary allows us to refer to the same device (such as `iosv-1`) with the appropriate IP address and prompt symbol. We can then use those values for the `expect()` method later on in the loop.

The output prints out the `show version | i V` output on the screen for each of the devices:

```
$ python3 chapter2_1.py
 b'show version | i VrnCisco IOS Software, IOSv
  Software (VIOS-ADVENTERPRISEK9-M), Version 15.6(2)T,
 RELEASE SOFTWARE (fc2)rnProcessor board ID
9MM4BI7B0DSWK40KV1IIRrn'
 b'show version | i VrnCisco IOS Software, IOSv
  Software (VIOS-ADVENTERPRISEK9-M), Version 15.6(2)T,
  RELEASE SOFTWARE (fc2)rn'
```

More Pexpect features

In this section, we will look at more Pexpect features that might come in handy when certain situations arise.

If you have a slow or fast link to your remote device, the default `expect()` method timeout is 30 seconds, which can be increased or decreased via the `timeout` argument:

```
>>> child.expect('Username', timeout=5)
```

You can choose to pass the command back to the user using the `interact()` method. This is useful when you just want to automate certain parts of the initial task:

```
>>> child.sendline('show version | i V')
19
>>> child.expect('iosv-1#')
0
>>> child.before
b'show version | i VrnCisco IOS Software, IOSv Software (VIOS-
ADVENTERPRISEK9-M), Version 15.6(2)T, RELEASE SOFTWARE (fc2)rnProcessor
board ID 9MM4BI7B0DSWK40KV1IIRrn'
>>> child.interact()
iosv-1#show run | i hostname
```

```
hostname iosv-1
iosv-1#exit
Connection closed by foreign host.
>>>
```

You can get a lot of information about the `child.spawn` object by printing it out in string format:

```
>>> str(child)
"<pexpect.pty_spawn.spawn object at 0x7fb01e29dba8>ncommand:
/usr/bin/telnetnargs: ['/usr/bin/telnet', '172.16.1.20']nsearcher:
Nonenbuffer (last 100 chars): b''nbefore (last 100 chars): b'NTERPRISEK9-
M), Version 15.6(2)T, RELEASE SOFTWARE (fc2)rnProcessor board ID
9MM4BI7B0DSWK40KV1IIRrn'nafter: b'iosv-1#'nmatch: <_sre.SRE_Match object;
span=(164, 171), match=b'iosv-1#'>nmatch_index: 0nexitstatus: 1nflag_eof:
Falsenpid: 2807nchild_fd: 5nclosed: Falsentimeout: 30ndelimiter: <class
'pexpect.exceptions.EOF'>nlogfile: Nonenlogfile_read: Nonenlogfile_send:
Nonenmaxread: 2000nignorecase: Falsensearchwindowsize:
Nonendelaybeforesend: 0.05ndelayafterclose: 0.1ndelayafterterminate: 0.1"
>>>
```

The most useful debug tool for Pexpect is to log the output in a file:

```
>>> child = pexpect.spawn('telnet 172.16.1.20')
>>> child.logfile = open('debug', 'wb')
```

Use `child.logfile = open('debug', 'w')` for Python 2. Python 3 uses byte strings by default. For more information on Pexpect features, check out https://pexpect.readthedocs.io/en/stable/api/index.html.

Pexpect and SSH

If you try to use the previous Telnet example and plug it into an SSH session instead, you might find yourself pretty frustrated with the experience. You always have to include the username in the session, answering the `ssh` new key question, and much more mundane tasks. There are many ways to make SSH sessions work, but luckily, Pexpect has a subclass called `pxssh`, which specializes in setting up SSH connections. The class adds methods for login, log out, and various tricky things to handle the different situations in the `ssh` login process. The procedures are mostly the same, with the exception of `login()` and `logout()`:

```
>>> from pexpect import pxssh
>>> child = pxssh.pxssh()
```

```
>>> child.login('172.16.1.20', 'cisco', 'cisco', auto_prompt_reset=False)
True
>>> child.sendline('show version | i V')
19
>>> child.expect('iosv-1#')
0
>>> child.before
b'show version | i VrnCisco IOS Software, IOSv Software (VIOS-
ADVENTERPRISEK9-M), Version 15.6(2)T, RELEASE SOFTWARE (fc2)rnProcessor
board ID 9MM4BI7B0DSWK40KV1IIRrn'
>>> child.logout()
>>>
```

Notice the `auto_prompt_reset=False` argument in the `login()` method. By default, `pxssh` uses the Shell prompt to synchronize the output. But since it uses the PS1 option for most of bash or CSH, they will error out on Cisco or other network devices.

Putting things together for Pexpect

As the final step, let's put everything you have learned so far about Pexpect into a script. Putting code into a script makes it easier to use in a production environment, as well as easier to share with your colleagues. We will write our second script, `chapter2_2.py`.

You can download the script from the book GitHub repository, https://github.com/PacktPublishing/Mastering-Python-Networking-second-edition, as well as looking at the output generated from the script as a result of the commands.

Refer to the following code:

```
#!/usr/bin/python3

import getpass
from pexpect import pxssh

devices = {'iosv-1': {'prompt': 'iosv-1#', 'ip': '172.16.1.20'},
'iosv-2': {'prompt': 'iosv-2#', 'ip': '172.16.1.21'}}
commands = ['term length 0', 'show version', 'show run']

username = input('Username: ')
password = getpass.getpass('Password: ')

# Starts the loop for devices
for device in devices.keys():
    outputFileName = device + '_output.txt'
    device_prompt = devices[device]['prompt']
```

```
        child = pxssh.pxssh()
        child.login(devices[device]['ip'], username.strip(),
password.strip(), auto_promp t_reset=False)
        # Starts the loop for commands and write to output
        with open(outputFileName, 'wb') as f:
            for command in commands:
                child.sendline(command)
                child.expect(device_prompt)
                f.write(child.before)

        child.logout()
```

The script further expands from our first Pexpect program with the following additional features:

- It uses SSH instead of Telnet
- It supports multiple commands instead of just one by making the commands into a list (line 8) and loops through the commands (starting at line 20)
- It prompts the user for their username and password instead of hardcoding them in the script
- It writes the output in two files, `iosv-1_output.txt`, and `ios-2_output.txt`, to be further analyzed

For Python 2, use `raw_input()` instead of `input()` for the username prompt. Also, use `w` for the file mode instead of `wb`.

The Python Paramiko library

Paramiko is a Python implementation of the SSHv2 protocol. Just like the `pxssh` subclass of Pexpect, Paramiko simplifies the SSHv2 interaction between the host and the remote device. Unlike `pxssh`, Paramiko focuses only on SSHv2 with no Telnet support. It also provides both client and server operations.

Paramiko is the low-level SSH client behind the high-level automation framework Ansible for its network modules. We will cover Ansible in later chapters. Let's take a look at the Paramiko library.

Installation of Paramiko

Installing Paramiko is pretty straightforward with Python `pip`. However, there is a hard dependency on the cryptography library. The library provides low-level, C-based encryption algorithms for the SSH protocol.

 The installation instruction for Windows, Mac, and other flavors of Linux can be found at https://cryptography.io/en/latest/installation/.

We will show the Paramiko installation of our Ubuntu 16.04 virtual machine in the following output. The following output shows the installation steps, as well as Paramiko successfully imported into the Python interactive prompt.

If you are using Python 2, please follow the steps below. We will try to import the library in the interactive prompt to make sure the library can be used:

```
sudo apt-get install build-essential libssl-dev libffi-dev python-dev
sudo pip install cryptography
sudo pip install paramiko
$ python
Python 2.7.12 (default, Nov 19 2016, 06:48:10)
[GCC 5.4.0 20160609] on linux2
Type "help", "copyright", "credits" or "license" for more information.
>>> import paramiko
>>> exit()
```

If you are using Python 3, please refer the following command-lines for installing the dependencies. After installation, we will import the library to make sure it is correctly installed:

```
sudo apt-get install build-essential libssl-dev libffi-dev python3-dev
sudo pip3 install cryptography
sudo pip3 install paramiko
$ python3
Python 3.5.2 (default, Nov 17 2016, 17:05:23)
[GCC 5.4.0 20160609] on linux
Type "help", "copyright", "credits" or "license" for more information.
>>> import paramiko
>>>
```

Paramiko overview

Let's look at a quick Paramiko example using the Python 3 interactive shell:

```
>>> import paramiko, time
>>> connection = paramiko.SSHClient()
>>> connection.set_missing_host_key_policy(paramiko.AutoAddPolicy())
>>> connection.connect('172.16.1.20', username='cisco', password='cisco',
look_for_keys=False, allow_agent=False)
>>> new_connection = connection.invoke_shell()
>>> output = new_connection.recv(5000)
>>> print(output)
b"rn*****************************************************************
***rn* IOSv is strictly limited to use for evaluation, demonstration and
IOS *rn* education. IOSv is provided as-is and is not supported by Cisco's
*rn* Technical Advisory Center. Any use or disclosure, in whole or in part,
*rn* of the IOSv Software or Documentation to any third party for any *rn*
purposes is expressly prohibited except as otherwise authorized by *rn*
Cisco in writing.
*rn*****************************************************************
**rniosv-1#"
>>> new_connection.send("show version | i Vn")
19
>>> time.sleep(3)
>>> output = new_connection.recv(5000)
>>> print(output)
b'show version | i VrnCisco IOS Software, IOSv Software (VIOS-
ADVENTERPRISEK9-M), Version 15.6(2)T, RELEASE SOFTWARE (fc2)rnProcessor
board ID 9MM4BI7B0DSWK40KV1IIRrniosv-1#'
>>> new_connection.close()
>>>
```

The `time.sleep()` function inserts a time delay to ensure that all the outputs were captured. This is particularly useful on a slower network connection or a busy device. This command is not required but is recommended depending on your situation.

Even if you are seeing the Paramiko operation for the first time, the beauty of Python and its clear syntax means that you can make a pretty good educated guess at what the program is trying to do:

```
>>> import paramiko
>>> connection = paramiko.SSHClient()
>>> connection.set_missing_host_key_policy(paramiko.AutoAddPolicy())
>>> connection.connect('172.16.1.20', username='cisco', password='cisco',
look_for_keys=False, allow_agent=False)
```

Low-Level Network Device Interactions

The first four lines create an instance of the `SSHClient` class from Paramiko. The next line sets the policy that the client should use when the SSH server's hostname; in this case, `iosv-1`, is not present in either the system host keys or the application's keys. In our scenario, we will automatically add the key to the application's `HostKeys` object. At this point, if you log on to the router, you will see the additional login session from Paramiko:

```
iosv-1#who
 Line User Host(s) Idle Location
*578 vty 0 cisco idle 00:00:00 172.16.1.1
 579 vty 1 cisco idle 00:01:30 172.16.1.173
 Interface User Mode Idle Peer Address
iosv-1#
```

The next few lines invoke a new interactive shell from the connection and a repeatable pattern of sending a command and retrieving the output. Finally, we close the connection.

Some readers who have used Paramiko before might be familiar with the `exec_command()` method instead of invoking a shell. Why do we need to invoke an interactive shell instead of using `exec_command()` directly? Unfortunately, `exec_command()` on Cisco IOS only allows a single command. Consider the following example with `exec_command()` for the connection:

```
>>> connection.connect('172.16.1.20', username='cisco', password='cisco',
look_for_keys=False, allow_agent=False)
>>> stdin, stdout, stderr = connection.exec_command('show version | i V')
>>> stdout.read()
b'Cisco IOS Software, IOSv Software (VIOS-ADVENTERPRISEK9-M), Version
15.6(2)T, RELEASE SOFTWARE (fc2)rnProcessor board ID
9MM4BI7B0DSWK40KV1IIRrn'
>>>
```

Everything works great; however, if you look at the number of sessions on the Cisco device, you will notice that the connection is dropped by the Cisco device without you closing the connection:

```
iosv-1#who
 Line User Host(s) Idle Location
*578 vty 0 cisco idle 00:00:00 172.16.1.1
 Interface User Mode Idle Peer Address
iosv-1#
```

Because the SSH session is no longer active, `exec_command()` will return an error if you want to send more commands to the remote device:

```
>>> stdin, stdout, stderr = connection.exec_command('show version | i V')
Traceback (most recent call last):
```

```
  File "<stdin>", line 1, in <module>
  File "/usr/local/lib/python3.5/dist-packages/paramiko/client.py", line 435, in exec_command
    chan = self._transport.open_session(timeout=timeout)
  File "/usr/local/lib/python3.5/dist-packages/paramiko/transport.py", line 711, in open_session
    timeout=timeout)
  File "/usr/local/lib/python3.5/dist-packages/paramiko/transport.py", line 795, in open_channel
    raise SSHException('SSH session not active')
paramiko.ssh_exception.SSHException: SSH session not active
>>>
```

> The Netmiko library by Kirk Byers is an open source Python library that simplifies SSH management to network devices. To read about it, check out this article, https://pynet.twb-tech.com/blog/automation/netmiko.html, and the source code, https://github.com/ktbyers/netmiko.

What would happen if you did not clear out the received buffer? The output would just keep on filling up the buffer and would overwrite it:

```
>>> new_connection.send("show version | i Vn")
19
>>> new_connection.send("show version | i Vn")
19
>>> new_connection.send("show version | i Vn")
19
>>> new_connection.recv(5000)
b'show version | i VrnCisco IOS Software, IOSv Software (VIOS-ADVENTERPRISEK9-M), Version 15.6(2)T, RELEASE SOFTWARE (fc2)rnProcessor board ID 9MM4BI7B0DSWK40KV1IIRrniosv-1#show version | i VrnCisco IOS Software, IOSv Software (VIOS-ADVENTERPRISEK9-M), Version 15.6(2)T, RELEASE SOFTWARE (fc2)rnProcessor board ID 9MM4BI7B0DSWK40KV1IIRrniosv-1#show version | i VrnCisco IOS Software, IOSv Software (VIOS-ADVENTERPRISEK9-M), Version 15.6(2)T, RELEASE SOFTWARE (fc2)rnProcessor board ID 9MM4BI7B0DSWK40KV1IIRrniosv-1#'
>>>
```

For consistency of the deterministic output, we will retrieve the output from the buffer each time we execute a command.

Our first Paramiko program

Our first program will use the same general structure as the Pexpect program we have put together. We will loop over a list of devices and commands while using Paramiko instead of Pexpect. This will give us a good compare and contrast of the differences between Paramiko and Pexpect.

If you have not done so already, you can download the code, `chapter2_3.py`, from the book's GitHub repository, `https://github.com/PacktPublishing/Mastering-Python-Networking-second-edition`. I will list the notable differences here:

```
devices = {'iosv-1': {'ip': '172.16.1.20'}, 'iosv-2': {'ip': '172.16.1.21'}}
```

We no longer need to match the device prompt using Paramiko; therefore, the device dictionary can be simplified:

```
commands = ['show version', 'show run']
```

There is no sendline equivalent in Paramiko; instead, we manually include the newline break in each of the commands:

```
def clear_buffer(connection):
    if connection.recv_ready():
        return connection.recv(max_buffer)
```

We include a new method to clear the buffer for sending commands, such as `terminal length 0` or `enable`, because we do not need the output for those commands. We simply want to clear the buffer and get to the execution prompt. This function will later be used in the loop, such as in line 25 of the script:

```
output = clear_buffer(new_connection)
```

The rest of the program should be pretty self-explanatory, similar to what we have seen in this chapter. The last thing I would like to point out is that since this is an interactive program, we place some buffer and wait for the command to be finished on the remote device before retrieving the output:

```
time.sleep(2)
```

After we clear the buffer, during the time between the execution of commands, we will wait two seconds. This will give the device adequate time to respond if it is busy.

More Paramiko features

We will look at Paramiko a bit later in the book, when we discuss Ansible, as Paramiko is the underlying transport for many of the network modules. In this section, we will take a look at some of the other features of Paramiko.

Paramiko for servers

Paramiko can be used to manage servers through SSHv2 as well. Let's look at an example of how we can use Paramiko to manage servers. We will use key-based authentication for the SSHv2 session.

In this example, I used another Ubuntu virtual machine on the same hypervisor as the destination server. You can also use a server on the VIRL simulator or an instance in one of the public cloud providers, such as Amazon AWS EC2.

We will generate a public-private key pair for our Paramiko host:

```
ssh-keygen -t rsa
```

This command, by default, will generate a public key named id_rsa.pub, as the public key under the user home directory ~/.ssh along with a private key named id_rsa. Treat the private key with the same attention as you would private passwords that you do not want to share with anybody else. You can think of the public key as a business card that identifies who you are. Using the private and public keys, the message will be encrypted by your private key locally and decrypted by the remote host using the public key. We should copy the public key to the remote host. In production, we can do this via out-of-band using a USB drive; in our lab, we can simply copy the public key to the remote host's ~/.ssh/authorized_keys file. Open up a Terminal window for the remote server, so you can paste in the public key.

Copy the content of ~/.ssh/id_rsa on your management host with Paramiko:

```
<Management Host with Pramiko>$ cat ~/.ssh/id_rsa.pub
ssh-rsa <your public key> echou@pythonicNeteng
```

Then, paste it to the remote host under the user directory; in this case, I am using echou for both sides:

```
<Remote Host>$ vim ~/.ssh/authorized_keys
ssh-rsa <your public key> echou@pythonicNeteng
```

Low-Level Network Device Interactions

You are now ready to use Paramiko to manage the remote host. Notice in this example that we will use the private key for authentication as well as the `exec_command()` method for sending commands:

```
Python 3.5.2 (default, Nov 17 2016, 17:05:23)
[GCC 5.4.0 20160609] on linux
Type "help", "copyright", "credits" or "license" for more information.
>>> import paramiko
>>> key = paramiko.RSAKey.from_private_key_file('/home/echou/.ssh/id_rsa')
>>> client = paramiko.SSHClient()
>>> client.set_missing_host_key_policy(paramiko.AutoAddPolicy())
>>> client.connect('192.168.199.182', username='echou', pkey=key)
>>> stdin, stdout, stderr = client.exec_command('ls -l')
>>> stdout.read()
b'total 44ndrwxr-xr-x 2 echou echou 4096 Jan 7 10:14 Desktopndrwxr-xr-x 2 echou echou 4096 Jan 7 10:14 Documentsndrwxr-xr-x 2 echou echou 4096 Jan 7 10:14 Downloadsn-rw-r--r-- 1 echou echou 8980 Jan 7 10:03 examples.desktopndrwxr-xr-x 2 echou echou 4096 Jan 7 10:14 Musicndrwxr-xr-x 2 echou echou 4096 Jan 7 10:14 Picturesndrwxr-xr-x 2 echou echou 4096 Jan 7 10:14 Publicndrwxr-xr-x 2 echou echou 4096 Jan 7 10:14 Templatesndrwxr-xr-x 2 echou echou 4096 Jan 7 10:14 Videosn'
>>> stdin, stdout, stderr = client.exec_command('pwd')
>>> stdout.read()
b'/home/echoun'
>>> client.close()
>>>
```

Notice that in the server example, we do not need to create an interactive session to execute multiple commands. You can now turn off password-based authentication in your remote host's SSHv2 configuration for more secure key-based authentication with automation enabled. Some network devices, such as Cumulus and Vyatta switches, also support key-based authentication.

Putting things together for Paramiko

We are almost at the end of the chapter. In this last section, let's make the Paramiko program more reusable. There is one downside of our existing script: we need to open up the script every time we want to add or delete a host, or whenever we need to change the commands we want to execute on the remote host. This is due to the fact that both the host and command information are statically entered inside of the script. Hardcoding the host and command has a higher chance of making mistakes. Besides, if you were to pass on the script to colleagues, they might not feel comfortable working in Python, Paramiko, or Linux.

By making both the host and command files be read in as parameters for the script, we can eliminate some of these concerns. Users (and a future you) can simply modify these text files when you need to make host or command changes.

We have incorporated the change in the script named `chapter2_4.py`.

Instead of hardcoding the commands, we broke the commands into a separate `commands.txt` file. Up to this point, we have been using show commands; in this example, we will make configuration changes. In particular, we will change the logging buffer size to 30000 bytes:

```
$ cat commands.txt
config t
logging buffered 30000
end
copy run start
```

The device's information is written into a `devices.json` file. We choose JSON format for the device's information because JSON data types can be easily translated into Python dictionary data types:

```
$ cat devices.json
{
    "iosv-1": {"ip": "172.16.1.20"},
    "iosv-2": {"ip": "172.16.1.21"}
}
```

In the script, we made the following changes:

```
with open('devices.json', 'r') as f:
    devices = json.load(f)

with open('commands.txt', 'r') as f:
    commands = [line for line in f.readlines()]
```

Here is an abbreviated output from the script execution:

```
$ python3 chapter2_4.py
Username: cisco
Password:
b'terminal length 0rniosv-2#config trnEnter configuration commands, one per line. End with CNTL/Z.rniosv-2(config)#'
b'logging buffered 30000rniosv-2(config)#'
...
```

Do a quick check to make sure the change has taken place in both `running-config` and `startup-config`:

```
iosv-1#sh run | i logging
logging buffered 30000
iosv-1#sh start | i logging
logging buffered 30000

iosv-2#sh run | i logging
logging buffered 30000
iosv-2#sh start | i logging
logging buffered 30000
```

Looking ahead

We have taken a pretty huge leap forward in this chapter as far as automating our network using Python is concerned. However, the method we have used feels like somewhat of a workaround for automation. We attempted to trick the remote devices into thinking they were interacting with a human on the other end.

Downsides of Pexpect and Paramiko compared to other tools

The biggest downside of our method so far is that the remote devices do not return structured data. They return data that is ideal to be fitted on a terminal to be interpreted by a human, not by a computer program. The human eye can easily interpret a space, while a computer only sees a return character.

We will take a look at a better way in the upcoming chapter. As a prelude to `Chapter 3`, *APIs and Intent-Driven Networking*, let's discuss the idea of idempotency.

Idempotent network device interaction

The term **idempotency** has different meanings, depending on its context. But in this book's context, the term means that when a client makes the same call to a remote device, the result should always be the same. I believe we can all agree that this is necessary. Imagine a scenario where each time you execute the same script, you get a different result back. I find that scenario very scary. How can you trust your script if that is the case? It would render our automation effort useless because we need to be prepared to handle different returns.

Since Pexpect and Paramiko are blasting out a series of commands interactively, the chance of having a non-idempotent interaction is higher. Going back to the fact that the return results needed to be screen scraped for useful elements, the risk of difference is much higher. Something on the remote end might have changed between the time we wrote the script and the time when the script is executed for the 100th time. For example, if the vendor makes a screen output change between releases without us updating the script, the script might break.

If we need to rely on the script for production, we need the script to be idempotent as much as possible.

Bad automation speeds bad things up

Bad automation allows you to poke yourself in the eye a lot faster, it is as simple as that. Computers are much faster at executing tasks than us human engineers. If we had the same set of operating procedures executed by a human versus a script, the script would finish faster than humans, sometimes without the benefit of having a solid feedback loop between procedures. The internet is full of horror stories of when someone pressed the *Enter* key and immediately regretted it.

We need to make sure the chances of bad automation scripts screwing things up are as small as possible. We all make mistakes; carefully test your script before any production work and small blast radius are two keys to making sure you can catch your mistake before it comes back and bites you.

Summary

In this chapter, we covered low-level ways to communicate directly with network devices. Without a way to programmatically communicate and make changes to network devices, there is no automation. We looked at two libraries in Python that allow us to manage devices that were meant to be managed by the CLI. Although useful, it is easy to see how the process can be somewhat fragile. This is mostly due to the fact that the network gears in question were meant to be managed by human beings and not computers.

In `Chapter 3`, *APIs and Intent-Driven Networking*, we will look at network devices supporting API and intent-driven networking.

3
APIs and Intent-Driven Networking

In Chapter 2, *Low-Level Network Device Interactions*, we looked at ways to interact with the network devices using Pexpect and Paramiko. Both of these tools use a persistent session that simulates a user typing in commands as if they are sitting in front of a Terminal. This works fine up to a point. It is easy enough to send commands over for execution on the device and capture the output. However, when the output becomes more than a few lines of characters, it becomes difficult for a computer program to interpret the output. The returned output from Pexpect and Paramiko is a series of characters meant to be read by a human being. The structure of the output consists of lines and spaces that are human-friendly but difficult to be understood by computer programs.

In order for our computer programs to automate many of the tasks we want to perform, we need to interpret the returned results and make follow-up actions based on the returned results. When we cannot accurately and predictably interpret the returned results, we cannot execute the next command with confidence.

Luckily, this problem was solved by the internet community. Imagine the difference between a computer and a human being when they are both reading a web page. The human sees words, pictures, and spaces interpreted by the browser; the computer sees raw HTML code, Unicode characters, and binary files. What happens when a website needs to become a web service for another computer? The same web resources need to accommodate both human clients and other computer programs. Doesn't this problem sound familiar to the one that we presented before? The answer is the **Application Program Interface** (**API**). It is important to note that an API is a concept and not a particular technology or framework, according to Wikipedia.

> *In computer programming, an **Application Programming Interface (API)** is a set of subroutine definitions, protocols, and tools for building application software. In general terms, it's a set of clearly defined methods of communication between various software components. A good API makes it easier to develop a computer program by providing all the building blocks, which are then put together by the programmer.*

In our use case, the set of clearly defined methods of communication would be between our Python program and the destination device. The APIs from our network devices provide a separate interface for the computer programs. The exact API implementation is vendor specific. One vendor will prefer XML over JSON, some might provide HTTPS as the underlying transport protocol, and others might provide Python libraries as wrappers. Despite the differences, the idea of an API remains the same: it is a separate communication method optimized for other computer programs.

In this chapter, we will look at the following topics:

- Treating infrastructure as code, intent-driven networking, and data modeling
- Cisco NX-API and the application-centric infrastructure
- Juniper NETCONF and PyEZ
- Arista eAPI and PyEAPI

Infrastructure as code

In a perfect world, network engineers and architects who design and manage networks should focus on what they want the network to achieve instead of the device-level interactions. In my first job as an intern for a local ISP, wide-eyed and excited, my first assignment was to install a router on a customer's site to turn up their fractional frame relay link (remember those?). How would I do that? I asked. I was handed a standard operating procedure for turning up frame relay links. I went to the customer site, blindly typed in the commands, and looked at the green lights flashing, then happily packed my bag and patted myself on the back for a job well done. As exciting as that first assignment was, I did not fully understand what I was doing. I was simply following instructions without thinking about the implication of the commands I was typing in. How would I troubleshoot something if the light was red instead of green? I think I would have called back to the office and cried for help (tears optional).

Of course, network engineering is not about typing in commands into a device, but it is about building a way that allows services to be delivered from one point to another with as little friction as possible. The commands we have to use and the output that we have to interpret are merely means to an end. In other words, we should be focused on our intent for the network. What we want our network to achieve is much more important than the command syntax we use to get the device to do what we want it to do. If we further extract that idea of describing our intent as lines of code, we can potentially describe our whole infrastructure as a particular state. The infrastructure will be described in lines of code with the necessary software or framework enforcing that state.

Intent-Driven Networking

Since the publication of the first edition of this book, the term **Intent-Based Networking** has seen an uptick in use after major network vendors chose to use it to describe their next-generation devices. In my opinion, **Intent-Driven Networking** is the idea of defining a state that the network should be in and having software code to enforce that state. As an example, if my goal is to block port 80 from being externally accessible, that is how I should declare it as the intention of the network. The underlying software will be responsible for knowing the syntax of configuring and applying the necessary access-list on the border router to achieve that goal. Of course, Intent-Driven Networking is an idea with no clear answer on the exact implementation. But the idea is simple and clear, I would hereby argue that we should focus as much on the intent of the network and abstract ourselves from the device-level interaction.

In using an API, it is my opinion that it gets us closer to a state of intent-driven networking. In short, because we abstract the layer of a specific command executed on our destination device, we focus on our intent instead of the specific commands. For example, going back to our `block port 80` access-list example, we might use access-list and access-group on a Cisco and filter-list on a Juniper. However, in using an API, our program can start asking the executor for their intent while masking what kind of physical device it is they are talking to. We can even use a higher-level declarative framework, such as Ansible, which we will cover in `Chapter 4`, *The Python Automation Framework – Ansible Basics*. But for now, let's focus on network APIs.

Screen scraping versus API structured output

Imagine a common scenario where we need to log into the network device and make sure all the interfaces on the devices are in an up/up state (both the status and the protocol are showing as up). For the human network engineers getting into a Cisco NX-OS device, it is simple enough to issue the `show IP interface brief` command in the Terminal to easily tell from the output which interface is up:

```
nx-osv-2# show ip int brief
IP Interface Status for VRF "default"(1)
Interface IP Address Interface Status
Lo0 192.168.0.2 protocol-up/link-up/admin-up
Eth2/1 10.0.0.6 protocol-up/link-up/admin-up
nx-osv-2#
```

The line break, white spaces, and the first line of the column title are easily distinguished from the human eye. In fact, they are there to help us line up, say, the IP addresses of each interface from line one to line two and three. If we were to put ourselves in the computer's position, all these spaces and line breaks only takes us away from the really important output, which is: which interfaces are in the up/up state? To illustrate this point, we can look at the Paramiko output for the same operation:

```
>>> new_connection.send('sh ip int briefn')
16
>>> output = new_connection.recv(5000)
>>> print(output)
b'sh ip int briefrrnIP Interface Status for VRF
"default"(1)rnInterface IP Address Interface
StatusrnLo0 192.168.0.2 protocol-up/link-up/admin-up
rnEth2/1 10.0.0.6 protocol-up/link-up/admin-up rnrnx-
osv-2# '
>>>
```

If we were to parse out that data, here is what I would do in a pseudo-code fashion (simplified representation of the code I would write):

1. Split each line via the line break.
2. I may or may not need the first line that contains the executed command of `show ip interface brief`. For now, I don't think I need it.
3. Take out everything on the second line up until the VRF, and save it in a variable as we want to know which VRF the output is showing.
4. For the rest of the lines, because we do not know how many interfaces there are, we will use a regular expression statement to search if the line starts with possible interfaces, such as `lo` for loopback and `Eth` for Ethernet interfaces.

5. We will need to split this line into three sections via space, each consisting of the name of the interface, IP address, and then the interface status.
6. The interface status will then be split further using the forward slash (/) to give us the protocol, link, and the admin status.

Whew, that is a lot of work just for something that a human being can tell at a glance! You might be able to optimize the code and the number of lines, but in general this is what we need to do when we need to screen scrap something that is somewhat unstructured. There are many downsides to this method, but some of the bigger problems that I can see are listed as follows:

- **Scalability**: We spent so much time on painstaking details to parse out the outputs from each command. It is hard to imagine how we can do this for the hundreds of commands that we typically run.
- **Predictability**: There is really no guarantee that the output stays the same between different software versions. If the output is changed ever so slightly, it might just render our hard-earned battle of information gathering useless.
- **Vendor and software lock-in**: Perhaps the biggest problem is that once we spend all this time parsing the output for this particular vendor and software version, in this case, Cisco NX-OS, we need to repeat this process for the next vendor that we pick. I don't know about you, but if I were to evaluate a new vendor, the new vendor is at a severe on-boarding disadvantage if I have to rewrite all the screen scrap code again.

Let's compare that with an output from an NX-API call for the same `show IP interface brief` command. We will go over the specifics of getting this output from the device later in this chapter, but what is important here is to compare the following output to the previous screen scraping output:

```
{
  "ins_api":{
    "outputs":{
      "output":{
        "body":{
          "TABLE_intf":[
            {
              "ROW_intf":{
                "admin-state":"up",
                "intf-name":"Lo0",
                "iod":84,
                "ip-disabled":"FALSE",
                "link-state":"up",
                "prefix":"192.168.0.2",
                "proto-state":"up"
```

```
            }
         },
         {
            "ROW_intf":{
            "admin-state":"up",
            "intf-name":"Eth2/1",
            "iod":36,
            "ip-disabled":"FALSE",
            "link-state":"up",
            "prefix":"10.0.0.6",
            "proto-state":"up"
            }
         }
         ],
          "TABLE_vrf":[
          {
          "ROW_vrf":{
          "vrf-name-out":"default"
          }
         },
         {
          "ROW_vrf":{
          "vrf-name-out":"default"
          }
         }
         ]
         },
         "code":"200",
         "input":"show ip int brief",
         "msg":"Success"
         }
         },
         "sid":"eoc",
         "type":"cli_show",
         "version":"1.2"
         }
      }
```

NX-API can return output in XML or JSON, and this is the JSON output that we are looking at. Right away, you can see the output is structured and can be mapped directly to the Python dictionary data structure. There is no parsing required—you can simply pick the key and retrieve the value associated with the key. You can also see from the output that there are various metadata in the output, such as the success or failure of the command. If the command fails, there will be a message telling the sender the reason for the failure. You no longer need to keep track of the command issued, because it is already returned to you in the `input` field. There is also other useful metadata in the output, such as the NX-API version.

Chapter 3

This type of exchange makes life easier for both vendors and operators. On the vendor side, they can easily transfer configuration and state information. They can add extra fields when the need to expose additional data arises using the same data structure. On the operator side, they can easily ingest the information and build their infrastructure around it. It is generally agreed on that automation is much needed and a good thing. The questions are usually centered on the format and structure of the automation. As you will see later in this chapter, there are many competing technologies under the umbrella of API. On the transport side alone, we have REST API, NETCONF, and RESTCONF, among others. Ultimately, the overall market might decide about the final data format in the future. In the meantime, each of us can form our own opinions and help drive the industry forward.

Data modeling for infrastructure as code

According to Wikipedia (`https://en.wikipedia.org/wiki/Data_model`), the definition for a data model is as follows:

> *A data model is an abstract model that organizes elements of data and standardizes how they relate to one another and to properties of the real-world entities. For instance, a data model may specify that the data element representing a car be composed of a number of other elements which, in turn, represent the color and size of the car and define its owner.*

The data modeling process can be illustrated in the following diagram:

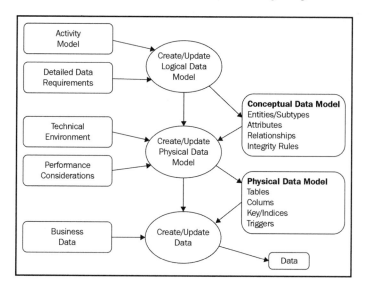

Data modeling process

[77]

When applied to the network, we can apply this concept as an abstract model that describes our network, be it a data center, campus, or global wide area network. If we take a closer look at a physical data center, a layer 2 Ethernet switch can be thought of as a device containing a table of MAC addresses mapped to each port. Our switch data model describes how the MAC address should be kept in a table, which includes the keys, additional characteristics (think of VLAN and private VLAN), and more. Similarly, we can move beyond devices and map the whole data center in a model. We can start with the number of devices in each of the access, distribution, and core layers, how they are connected, and how they should behave in a production environment. For example, if we have a fat-tree network, how many links should each of the spine routers have, how many routes they should contain, and how many next-hops should each of the prefixes have? These characteristics can be mapped out in a format that can be referenced against the ideal state that we should always check against.

One of the relatively new network data modeling languages that is gaining traction is **Yet Another Next Generation** (**YANG**) (despite common belief, some of the IETF workgroups do have a sense of humor). It was first published in RFC 6020 in 2010, and has since gained traction among vendors and operators. At the time of writing, the support for YANG has varied greatly from vendors to platforms. The adaptation rate in production is therefore relatively low. However, it is a technology worth keeping an eye out for.

The Cisco API and ACI

Cisco Systems, the 800-pound gorilla in the networking space, have not missed out on the trend of network automation. In their push for network automation, they have made various in-house developments, product enhancements, partnerships, as well as many external acquisitions. However, with product lines spanning routers, switches, firewalls, servers (unified computing), wireless, the collaboration software and hardware, and analytic software, to name a few, it is hard to know where to start.

Since this book focuses on Python and networking, we will scope this section to the main networking products. In particular, we will cover the following:

- Nexus product automation with NX-API
- Cisco NETCONF and YANG examples
- The Cisco application-centric infrastructure for the data center
- The Cisco application-centric infrastructure for the enterprise

Chapter 3

For the NX-API and NETCONF examples here, we can either use the Cisco DevNet always-on lab devices or locally run Cisco VIRL. Since ACI is a separate product and is licensed with the physical switches for the following ACI examples, I would recommend using the DevNet labs to get an understanding of the tools. If you are one of the lucky engineers who has a private ACI lab that you can use, please feel free to use it for the relevant examples.

We will use the similar lab topology as we did in Chapter 2, *Low-Level Network Device Interactions*, with the exception of one of the devices running **nx-osv**:

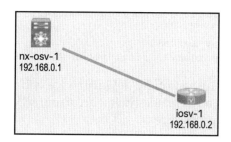

Lab topology

Let's take a look at NX-API.

Cisco NX-API

Nexus is Cisco's primary product line of data center switches. The NX-API (http://www.cisco.com/c/en/us/td/docs/switches/datacenter/nexus9000/sw/6-x/programmability/guide/b_Cisco_Nexus_9000_Series_NX-OS_Programmability_Guide/b_Cisco_Nexus_9000_Series_NX-OS_Programmability_Guide_chapter_011.html) allows the engineer to interact with the switch outside of the device via a variety of transports including SSH, HTTP, and HTTPS.

Lab software installation and device preparation

Here are the Ubuntu packages that we will install. You may already have some of the packages such pip and git:

```
$ sudo apt-get install -y python3-dev libxml2-dev libxslt1-dev libffi-dev libssl-dev zlib1g-dev python3-pip git python3-requests
```

APIs and Intent-Driven Networking

 If you are using Python 2, use the following packages instead: `sudo apt-get install -y python-dev libxml2-dev libxslt1-dev libffi-dev libssl-dev zlib1g-dev python-pip git python-requests`.

The `ncclient` (https://github.com/ncclient/ncclient) library is a Python library for NETCONF clients. We will install this from the GitHub repository so that we can install the latest version:

```
$ git clone https://github.com/ncclient/ncclient
$ cd ncclient/
$ sudo python3 setup.py install
$ sudo python setup.py install #for Python 2
```

NX-API on Nexus devices is turned off by default, so we will need to turn it on. We can either use the user that is already created (if you are using VIRL auto-config), or create a new user for the NETCONF procedures:

```
feature nxapi
username cisco password 5 $1$Nk7ZkwH0$fyiRmMMfIheqE3BqvcL0C1 role network-operator
username cisco role network-admin
username cisco passphrase lifetime 99999 warntime 14 gracetime 3
```

For our lab, we will turn on both HTTP and the sandbox configuration, as they should be turned off in production:

```
nx-osv-2(config)# nxapi http port 80
nx-osv-2(config)# nxapi sandbox
```

We are now ready to look at our first NX-API example.

NX-API examples

NX-API sandbox is a great way to play around with various commands, data formats, and even copy the Python script directly from the web page. In the last step, we turned it on for learning purposes. It should be turned off in production. Let's launch a web browser and take a look at the various message formats, requests, and responses based on the CLI commands that we are already familiar with:

Chapter 3

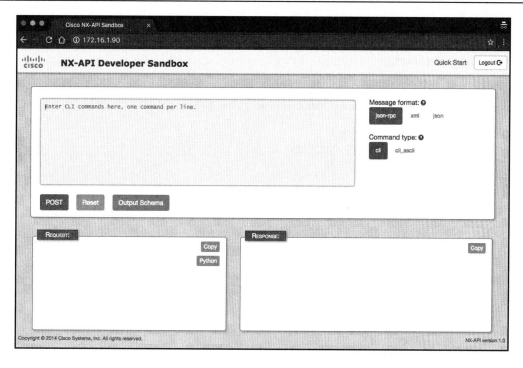

In the following example, I have selected JSON-RPC and the CLI command type for the show version command:

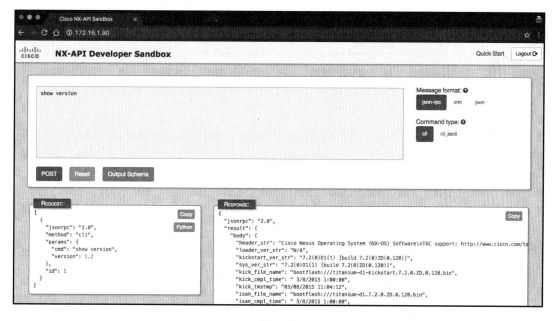

[81]

APIs and Intent-Driven Networking

The sandbox comes in handy if you are unsure about the supportability of the message format, or if you have questions about the response data field keys for the value you want to retrieve in your code.

In our first example, we are just going to connect to the Nexus device and print out the capabilities exchanged when the connection was first made:

```python
#!/usr/bin/env python3
from ncclient import manager
conn = manager.connect(
        host='172.16.1.90',
        port=22,
        username='cisco',
        password='cisco',
        hostkey_verify=False,
        device_params={'name': 'nexus'},
        look_for_keys=False)
for value in conn.server_capabilities:
    print(value)
conn.close_session()
```

The connection parameters of the host, port, username, and password are pretty self-explanatory. The device parameter specifies the kind of device the client is connecting to. We will see a different response in the Juniper NETCONF sections when using the ncclient library. The `hostkey_verify` bypasses the `known_host` requirement for SSH; if not, the host needs to be listed in the `~/.ssh/known_hosts` file. The `look_for_keys` option disables public-private key authentication, but uses a username and password for authentication.

If you run into an issue with https://github.com/paramiko/paramiko/issues/748 with Python 3 and Paramiko, please feel free to use Python 2. Hopefully, by the time you read this section, the issue is already fixed.

The output will show the XML and NETCONF supported features by this version of NX-OS:

```
$ python cisco_nxapi_1.py
urn:ietf:params:netconf:capability:writable-running:1.0
urn:ietf:params:netconf:capability:rollback-on-error:1.0
urn:ietf:params:netconf:capability:validate:1.0
urn:ietf:params:netconf:capability:url:1.0?scheme=file
urn:ietf:params:netconf:base:1.0
urn:ietf:params:netconf:capability:candidate:1.0
urn:ietf:params:netconf:capability:confirmed-commit:1.0
urn:ietf:params:xml:ns:netconf:base:1.0
```

Chapter 3

Using ncclient and NETCONF over SSH is great because it gets us closer to the native implementation and syntax. We will use the same library later on in this book. For NX-API, it might be easier to deal with HTTPS and JSON-RPC. In the earlier screenshot of **NX-API Developer Sandbox**, if you noticed, in the **Request** box, there is a box labeled **Python**. If you click on it, you will be able to get an automatically converted Python script based on the request library.

The following script uses an external Python library named requests. requests is a very popular, self-proclaimed HTTP for the human library used by companies like Amazon, Google, NSA, and more. You can find more information about it on the official site (http://docs.python-requests.org/en/master/).

For the show version example, the following Python script is automatically generated for you. I am pasting in the output without any modification:

```
"""
NX-API-BOT
"""
import requests
import json

"""
Modify these please
"""
url='http://YOURIP/ins'
switchuser='USERID'
switchpassword='PASSWORD'

myheaders={'content-type':'application/json-rpc'}
payload=[
  {
    "jsonrpc": "2.0",
    "method": "cli",
    "params": {
      "cmd": "show version",
      "version": 1.2
    },
    "id": 1
  }
]
response = requests.post(url,data=json.dumps(payload),
headers=myheaders,auth=(switchuser,switchpassword)).json()
```

APIs and Intent-Driven Networking

In the `cisco_nxapi_2.py` script, you will see that I have only modified the URL, username, and password of the preceding file. The output was parsed to include only the software version. Here is the output:

```
$ python3 cisco_nxapi_2.py
7.2(0)D1(1) [build 7.2(0)ZD(0.120)]
```

The best part about using this method is that the same overall syntax structure works with both configuration commands as well as show commands. This is illustrated in the `cisco_nxapi_3.py` file. For multiline configuration, you can use the ID field to specify the order of operations. In `cisco_nxapi_4.py`, the following payload was listed for changing the description of the interface Ethernet 2/12 in the interface configuration mode:

```
{
  "jsonrpc": "2.0",
  "method": "cli",
  "params": {
    "cmd": "interface ethernet 2/12",
    "version": 1.2
  },
  "id": 1
},
{
  "jsonrpc": "2.0",
  "method": "cli",
  "params": {
    "cmd": "description foo-bar",
    "version": 1.2
  },
  "id": 2
},
{
  "jsonrpc": "2.0",
  "method": "cli",
  "params": {
    "cmd": "end",
    "version": 1.2
  },
  "id": 3
},
{
  "jsonrpc": "2.0",
  "method": "cli",
  "params": {
    "cmd": "copy run start",
    "version": 1.2
  },
```

```
            "id": 4
    }
]
```

We can verify the result of the previous configuration script by looking at the running-configuration of the Nexus device:

```
hostname nx-osv-1-new
...
interface Ethernet2/12
description foo-bar
shutdown
no switchport
mac-address 0000.0000.002f
```

In the next section, we will look at some examples for Cisco NETCONF and the YANG model.

The Cisco and YANG models

Earlier in this chapter, we looked at the possibility of expressing the network by using the data modeling language YANG. Let's look into it a little bit more with examples.

First off, we should know that the YANG model only defines the type of data sent over the NETCONF protocol without dictating what the data should be. Secondly, it is worth pointing out that NETCONF exists as a standalone protocol, as we saw in the NX-API section. YANG, being relatively new, has a spotty supportability across vendors and product lines. For example, if we run the same capability exchange script that we have used before for a Cisco 1000v running IOS-XE, this is what we will see:

```
urn:cisco:params:xml:ns:yang:cisco-virtual-service?module=cisco-
virtual-service&revision=2015-04-09
http://tail-f.com/ns/mibs/SNMP-NOTIFICATION-MIB/200210140000Z?
module=SNMP-NOTIFICATION-MIB&revision=2002-10-14
urn:ietf:params:xml:ns:yang:iana-crypt-hash?module=iana-crypt-
hash&revision=2014-04-04&features=crypt-hash-sha-512,crypt-hash-
sha-256,crypt-hash-md5
urn:ietf:params:xml:ns:yang:smiv2:TUNNEL-MIB?module=TUNNEL-
MIB&revision=2005-05-16
urn:ietf:params:xml:ns:yang:smiv2:CISCO-IP-URPF-MIB?module=CISCO-
IP-URPF-MIB&revision=2011-12-29
urn:ietf:params:xml:ns:yang:smiv2:ENTITY-STATE-MIB?module=ENTITY-
STATE-MIB&revision=2005-11-22
urn:ietf:params:xml:ns:yang:smiv2:IANAifType-MIB?module=IANAifType-
MIB&revision=2006-03-31
```

```
<omitted>
```

Compare this to the output that we saw for NX-OS. Clearly, IOS-XE supports the YANG model features more than NX-OS. Industry-wide, network data modeling when supported, is clearly something that can be used across your devices, which is beneficial for network automation. However, given the uneven support of vendors and products, it is not yet mature enough to be used exclusively for the production network, in my opinion. For this book, I have included a script called `cisco_yang_1.py` that shows how to parse out the NETCONF XML output with YANG filters called `urn:ietf:params:xml:ns:yang:ietf-interfaces` as a starting point to see the existing tag overlay.

You can check the latest vendor support on the YANG GitHub project page (https://github.com/YangModels/yang/tree/master/vendor).

The Cisco ACI

The Cisco **Application Centric Infrastructure** (**ACI**) is meant to provide a centralized approach to all of the network components. In the data center context, it means that the centralized controller is aware of and manages the spine, leaf, and top of rack switches, as well as all the network service functions. This can be done through GUI, CLI, or API. Some might argue that the ACI is Cisco's answer to the broader controller-based software-defined networking.

One of the somewhat confusing points for ACI is the difference between ACI and APIC-EM. In short, ACI focuses on data center operations while APIC-EM focuses on enterprise modules. Both offer a centralized view and control of the network components, but each has its own focus and share of tool sets. For example, it is rare to see any major data center deploy a customer-facing wireless infrastructure, but a wireless network is a crucial part of enterprises today. Another example would be the different approaches to network security. While security is important in any network, in the data center environment, lots of security policies are pushed to the edge node on the server for scalability. In enterprise security, policies are somewhat shared between the network devices and servers.

Unlike NETCONF RPC, ACI API follows the REST model to use the HTTP verb (`GET`, `POST`, `DELETE`) to specify the operation that's intended.

Chapter 3

We can look at the `cisco_apic_em_1.py` file, which is a modified version of the Cisco sample code on `lab2-1-get-network-device-list.py` (https://github.com/CiscoDevNet/apicem-1.3-LL-sample-codes/blob/master/basic-labs/lab2-1-get-network-device-list.py).

The abbreviated version without comments and spaces are listed in the following section.

The first function named `getTicket()` uses HTTPS POST on the controller with the path of `/api/v1/ticket` with a username and password embedded in the header. This function will return the parsed response for a ticket that is only valid for a limited time:

```
def getTicket():
    url = "https://" + controller + "/api/v1/ticket"
    payload = {"username":"usernae","password":"password"}
    header = {"content-type": "application/json"}
    response= requests.post(url,data=json.dumps(payload), headers=header, verify=False)
    r_json=response.json()
    ticket = r_json["response"]["serviceTicket"]
    return ticket
```

The second function then calls another path called `/api/v1/network-devices` with the newly acquired ticket embedded in the header, then parses the results:

```
url = "https://" + controller + "/api/v1/network-device"
header = {"content-type": "application/json", "X-Auth-Token":ticket}
```

This is a pretty common workflow for API interactions. The client will authenticate itself with the server in the first request and receive a time-based token. This token will be used in subsequent requests and will be served as a proof of authentication.

The output displays both the raw JSON response output as well as a parsed table. A partial output when executed against a DevNet lab controller is shown here:

```
Network Devices =
{
  "version": "1.0",
  "response": [
    {
    "reachabilityStatus": "Unreachable",
    "id": "8dbd8068-1091-4cde-8cf5-d1b58dc5c9c7",
    "platformId": "WS-C2960C-8PC-L",
<omitted>
    "lineCardId": null,
    "family": "Wireless Controller",
    "interfaceCount": "12",
```

[87]

```
        "upTime": "497 days, 2:27:52.95"
      }
    ]
  }
8dbd8068-1091-4cde-8cf5-d1b58dc5c9c7 Cisco Catalyst 2960-C Series
  Switches
cd6d9b24-839b-4d58-adfe-3fdf781e1782 Cisco 3500I Series Unified
Access Points
<omitted>
55450140-de19-47b5-ae80-bfd741b23fd9 Cisco 4400 Series Integrated
Services Routers
ae19cd21-1b26-4f58-8ccd-d265deabb6c3 Cisco 5500 Series Wireless LAN
Controllers
```

As you can see, we only query a single controller device, but we are able to get a high-level view of all the network devices that the controller is aware of. In our output, the Catalyst 2960-C switch, 3500 Access Points, 4400 ISR router, and 5500 Wireless Controller can all be explored further. The downside is, of course, that the ACI controller only supports Cisco devices at this time.

The Python API for Juniper networks

Juniper networks have always been a favorite among the service provider crowd. If we take a step back and look at the service provider vertical, it would make sense that automating network equipment is on the top of their list of requirements. Before the dawn of cloud-scale data centers, service providers were the ones with the most network equipment. A typical enterprise network might have a few redundant internet connections at the corporate headquarter with a few hub-and-spoke remote sites connected back to the HQ using the service provider's private MPLS network. To a service provider, they are the ones who need to build, provision, manage, and troubleshoot the connections and the underlying networks. They make their money by selling the bandwidth along with value-added managed services. It would make sense for the service providers to invest in automation to use the least amount of engineering hours to keep the network humming along. In their use case, network automation is the key to their competitive advantage.

In my opinion, the difference between a service provider's network needs compared to a cloud data center is that, traditionally, service providers aggregate more services into a single device. A good example would be **Multiprotocol Label Switching** (**MPLS**) that almost all major service providers provide but rarely adapt in the enterprise or data center networks. Juniper, as they have been very successful, has identified this need and excel at fulfilling the service provider requirements of automating. Let's take a look at some of Juniper's automation APIs.

Chapter 3

Juniper and NETCONF

The **Network Configuration Protocol** (**NETCONF**) is an IETF standard, which was first published in 2006 as RFC 4741 and later revised in RFC 6241. Juniper networks contributed heavily to both of the RFC standards. In fact, Juniper was the sole author for RFC 4741. It makes sense that Juniper devices fully support NETCONF, and it serves as the underlying layer for most of its automation tools and frameworks. Some of the main characteristics of NETCONF include the following:

1. It uses **Extensible Markup Language** (**XML**) for data encoding.
2. It uses **Remote Procedure Calls** (**RPC**), therefore in the case of HTTP(s) as the transport, the URL endpoint is identical while the operation intended is specified in the body of the request.
3. It is conceptually based on layers from top to bottom. The layers include the content, operations, messages, and transport:

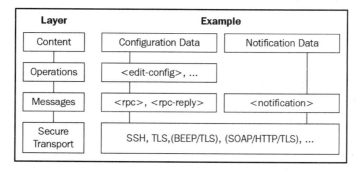

NETCONF model

Juniper networks provide an extensive NETCONF XML management protocol developer guide (https://www.juniper.net/techpubs/en_US/junos13.2/information-products/pathway-pages/netconf-guide/netconf.html#overview) in its technical library. Let's take a look at its usage.

Device preparation

In order to start using NETCONF, let's create a separate user as well as turn on the required services:

```
set system login user netconf uid 2001
set system login user netconf class super-user
set system login user netconf authentication encrypted-password
```

APIs and Intent-Driven Networking

```
    "$1$0EkA.XVf$cm80A0GC2dgSWJIYWv7Pt1"
set system services ssh
set system services telnet
set system services netconf ssh port 830
```

For the Juniper device lab, I am using an older, unsupported platform called **Juniper Olive**. It is solely used for lab purposes. You can use your favorite search engine to find out some interesting facts and history about Juniper Olive.

On the Juniper device, you can always take a look at the configuration either in a flat file or in XML format. The `flat` file comes in handy when you need to specify a one-liner command to make configuration changes:

```
netconf@foo> show configuration | display set
set version 12.1R1.9
set system host-name foo
set system domain-name bar
<omitted>
```

The XML format comes in handy at times when you need to see the XML structure of the configuration:

```
netconf@foo> show configuration | display xml
<rpc-reply xmlns:junos="http://xml.juniper.net/junos/12.1R1/junos">
  <configuration junos:commit-seconds="1485561328" junos:commit-
localtime="2017-01-27 23:55:28 UTC" junos:commit-user="netconf">
  <version>12.1R1.9</version>
  <system>
  <host-name>foo</host-name>
  <domain-name>bar</domain-name>
```

We have installed the necessary Linux libraries and the ncclient Python library in the Cisco section. If you have not done so, refer back to that section and install the necessary packages.

We are now ready to look at our first Juniper NETCONF example.

Juniper NETCONF examples

We will use a pretty straightforward example to execute `show version`. We will name this file `junos_netconf_1.py`:

```
#!/usr/bin/env python3

from ncclient import manager

conn = manager.connect(
    host='192.168.24.252',
    port='830',
    username='netconf',
    password='juniper!',
    timeout=10,
    device_params={'name':'junos'},
    hostkey_verify=False)

result = conn.command('show version', format='text')
print(result)
conn.close_session()
```

All the fields in the script should be pretty self-explanatory, with the exception of `device_params`. Starting with ncclient 0.4.1, the device handler was added to specify different vendors or platforms. For example, the name can be juniper, CSR, Nexus, or Huawei. We also added `hostkey_verify=False` because we are using a self-signed certificate from the Juniper device.

The returned output is `rpc-reply` encoded in XML with an `output` element:

```
<rpc-reply message-id="urn:uuid:7d9280eb-1384-45fe-be48-b7cd14ccf2b7">
<output>
Hostname: foo
Model: olive
JUNOS Base OS boot [12.1R1.9]
JUNOS Base OS Software Suite [12.1R1.9]
<omitted>
JUNOS Runtime Software Suite [12.1R1.9]
JUNOS Routing Software Suite [12.1R1.9]
</output>
</rpc-reply>
```

We can parse the XML output to just include the output text:

```
print(result.xpath('output')[0].text)
```

APIs and Intent-Driven Networking

In `junos_netconf_2.py`, we will make configuration changes to the device. We will start with some new imports for constructing new XML elements and the connection manager object:

```
#!/usr/bin/env python3

from ncclient import manager
from ncclient.xml_ import new_ele, sub_ele

conn = manager.connect(host='192.168.24.252', port='830',
username='netconf' , password='juniper!', timeout=10,
device_params={'name':'junos'}, hostkey_verify=False)
```

We will lock the configuration and make configuration changes:

```
# lock configuration and make configuration changes
conn.lock()

# build configuration
config = new_ele('system')
sub_ele(config, 'host-name').text = 'master'
sub_ele(config, 'domain-name').text = 'python'
```

Under the build configuration section, we create a new element of `system` with subelements of `host-namre` and `domain-name`. If you were wondering about the hierarchy structure, you can see from the XML display that the node structure with `system` is the parent of `host-name` and `domain-name`:

```
<system>
    <host-name>foo</host-name>
    <domain-name>bar</domain-name>
...
</system>
```

After the configuration is built, the script will push the configuration and commit the configuration changes. These are the normal best practice steps (lock, configure, unlock, commit) for Juniper configuration changes:

```
# send, validate, and commit config
conn.load_configuration(config=config)
conn.validate()
commit_config = conn.commit()
print(commit_config.tostring)

# unlock config
conn.unlock()
```

Chapter 3

```
    # close session
    conn.close_session()
```

Overall, the NETCONF steps map pretty well to what you would have done in the CLI steps. Please take a look at the `junos_netconf_3.py` script for a more reusable code. The following example combines the step-by-step example with a few Python functions:

```
# make a connection object
def connect(host, port, user, password):
    connection = manager.connect(host=host, port=port, username=user,
                password=password, timeout=10, device_params={'name':'junos'},
                hostkey_verify=False)
    return connection

# execute show commands
def show_cmds(conn, cmd):
    result = conn.command(cmd, format='text')
    return result

# push out configuration
def config_cmds(conn, config):
    conn.lock()
    conn.load_configuration(config=config)
    commit_config = conn.commit()
    return commit_config.tostring
```

This file can be executed by itself, or it can be imported to be used by other Python scripts.

Juniper also provides a Python library to be used with their devices called PyEZ. We will take a look at a few examples of using the library in the following section.

Juniper PyEZ for developers

PyEZ is a high-level Python implementation that integrates better with your existing Python code. By utilizing the Python API, you can perform common operation and configuration tasks without the extensive knowledge of the Junos CLI.

Juniper maintains a comprehensive Junos PyEZ developer guide at https://www.juniper.net/techpubs/en_US/junos-pyez1.0/information-products/pathway-pages/junos-pyez-developer-guide.html#configuration on their technical library. If you are interested in using PyEZ, I would highly recommend at least a glance through the various topics in the guide.

[93]

Installation and preparation

The installation instructions for each of the operating systems can be found on the *Installing Junos PyEZ* (https://www.juniper.net/techpubs/en_US/junos-pyez1.0/topics/task/installation/junos-pyez-server-installing.html) page. We will show the installation instructions for Ubuntu 16.04.

The following are some dependency packages, many of which should already be on the host from running previous examples:

```
$ sudo apt-get install -y python3-pip python3-dev libxml2-dev libxslt1-dev libssl-dev libffi-dev
```

`PyEZ` packages can be installed via pip. Here, I have installed for both Python 3 and Python 2:

```
$ sudo pip3 install junos-eznc
$ sudo pip install junos-eznc
```

On the Juniper device, NETCONF needs to be configured as the underlying XML API for PyEZ:

```
set system services netconf ssh port 830
```

For user authentication, we can either use password authentication or an SSH key pair. Creating the local user is straightforward:

```
set system login user netconf uid 2001
set system login user netconf class super-user
set system login user netconf authentication encrypted-password "$1$0EkA.XVf$cm80A0GC2dgSWJIYWv7Pt1"
```

For the `ssh` key authentication, first, generate the key pair on your host:

```
$ ssh-keygen -t rsa
```

By default, the public key will be called `id_rsa.pub` under `~/.ssh/`, while the private key will be named `id_rsa` under the same directory. Treat the private key like a password that you never share. The public key can be freely distributed. In our use case, we will move the public key to the `/tmp` directory and enable the Python 3 HTTP server module to create a reachable URL:

```
$ mv ~/.ssh/id_rsa.pub /tmp
$ cd /tmp
$ python3 -m http.server
Serving HTTP on 0.0.0.0 port 8000 ...
```

Chapter 3

 For Python 2, use `python -m SimpleHTTPServer` instead.

From the Juniper device, we can create the user and associate the public key by downloading the public key from the Python 3 web server:

```
netconf@foo# set system login user echou class super-user authentication load-key-file http://192.168.24.164:8000/id_rsa.pub
/var/home/netconf/...transferring.file.......100% of 394 B 2482 kBps
```

Now, if we try to `ssh` with the private key from the management station, the user will be automatically authenticated:

```
$ ssh -i ~/.ssh/id_rsa 192.168.24.252
--- JUNOS 12.1R1.9 built 2012-03-24 12:52:33 UTC
echou@foo>
```

Let's make sure that both of the authentication methods work with PyEZ. Let's try the username and password combination:

```
Python 3.5.2 (default, Nov 17 2016, 17:05:23)
[GCC 5.4.0 20160609] on linux
Type "help", "copyright", "credits" or "license" for more information.
>>> from jnpr.junos import Device
>>> dev = Device(host='192.168.24.252', user='netconf', password='juniper!')
>>> dev.open()
Device(192.168.24.252)
>>> dev.facts
{'serialnumber': '', 'personality': 'UNKNOWN', 'model': 'olive', 'ifd_style': 'CLASSIC', '2RE': False, 'HOME': '/var/home/juniper', 'version_info': junos.version_info(major=(12, 1), type=R, minor=1, build=9), 'switch_style': 'NONE', 'fqdn': 'foo.bar', 'hostname': 'foo', 'version': '12.1R1.9', 'domain': 'bar', 'vc_capable': False}
>>> dev.close()
```

We can also try to use the SSH key authentication:

```
>>> from jnpr.junos import Device
>>> dev1 = Device(host='192.168.24.252', user='echou', ssh_private_key_file='/home/echou/.ssh/id_rsa')
>>> dev1.open()
Device(192.168.24.252)
>>> dev1.facts
{'HOME': '/var/home/echou', 'model': 'olive', 'hostname': 'foo',
```

[95]

```
'switch_style': 'NONE', 'personality': 'UNKNOWN', '2RE': False, 'domain':
'bar', 'vc_capable': False, 'version': '12.1R1.9', 'serialnumber': '',
'fqdn': 'foo.bar', 'ifd_style': 'CLASSIC', 'version_info':
junos.version_info(major=(12, 1), type=R, minor=1, build=9)}
>>> dev1.close()
```

Great! We are now ready to look at some examples for PyEZ.

PyEZ examples

In the previous interactive prompt, we already saw that when the device connects, the object automatically retrieves a few facts about the device. In our first example, `junos_pyez_1.py`, we were connecting to the device and executing an RPC call for `show interface em1`:

```
#!/usr/bin/env python3
from jnpr.junos import Device
import xml.etree.ElementTree as ET
import pprint

dev = Device(host='192.168.24.252', user='juniper', passwd='juniper!')

try:
    dev.open()
except Exception as err:
    print(err)
    sys.exit(1)

result = dev.rpc.get_interface_information(interface_name='em1', terse=True)
pprint.pprint(ET.tostring(result))

dev.close()
```

The device class has an `rpc` property that includes all operational commands. This is pretty awesome because there is no slippage between what we can do in CLI versus API. The catch is that we need to find out the `xml rpc` element tag. In our first example, how do we know `show interface em1` equates to `get_interface_information`? We have three ways of finding out this information:

1. We can reference the *Junos XML API Operational Developer Reference*
2. We can use the CLI and display the XML RPC equivalent and replace the dash (-) between the words with an underscore (_)

Chapter 3

3. We can also do this programmatically by using the PyEZ library

I typically use the second option to get the output directly:

```
netconf@foo> show interfaces em1 | display xml rpc
<rpc-reply xmlns:junos="http://xml.juniper.net/junos/12.1R1/junos">
 <rpc>
 <get-interface-information>
 <interface-name>em1</interface-name>
 </get-interface-information>
 </rpc>
 <cli>
 <banner></banner>
 </cli>
</rpc-reply>
```

Here is an example of using PyEZ programmatically (the third option):

```
>>> dev1.display_xml_rpc('show interfaces em1', format='text')
'<get-interface-information>n <interface-name>em1</interface-name>n</get-interface-information>n'
```

Of course, we will need to make configuration changes as well. In the `junos_pyez_2.py` configuration example, we will import an additional `Config()` method from PyEZ:

```
#!/usr/bin/env python3
from jnpr.junos import Device
from jnpr.junos.utils.config import Config
```

We will utilize the same block for connecting to a device:

```
dev = Device(host='192.168.24.252', user='juniper',
passwd='juniper!')

try:
    dev.open()
except Exception as err:
    print(err)
    sys.exit(1)
```

The new `Config()` method will load the XML data and make the configuration changes:

```
config_change = """
<system>
  <host-name>master</host-name>
  <domain-name>python</domain-name>
</system>
"""
```

[97]

```
cu = Config(dev)
cu.lock()
cu.load(config_change)
cu.commit()
cu.unlock()

dev.close()
```

The PyEZ examples are simple by design. Hopefully, they demonstrate the ways you can leverage PyEZ for your Junos automation needs.

The Arista Python API

Arista Networks have always been focused on large-scale data center networks. In its corporate profile page (https://www.arista.com/en/company/company-overview), it is stated as follows:

> *"Arista Networks was founded to pioneer and deliver software-driven cloud networking solutions for large data center storage and computing environments."*

Notice that the statement specifically called out **large data centers**, which we already know are exploded with servers, databases, and, yes, network equipment. It makes sense that automation has always been one of Arista's leading features. In fact, they have a Linux underpin behind their operating system, allowing many added benefits such as Linux commands and a built-in Python interpreter.

Like other vendors, you can interact with Arista devices directly via eAPI, or you can choose to leverage their `Python` library. We will see examples of both. We will also look at Arista's integration with the Ansible framework in later chapters.

Arista eAPI management

Arista's eAPI was first introduced in EOS 4.12 a few years ago. It transports a list of show or configuration commands over HTTP or HTTPS and responds back in JSON. An important distinction is that it is a **Remote Procedure Call** (**RPC**) and **JSON-RPC**, instead of a pure RESTFul API that's served over HTTP or HTTPS. For our intents and purposes, the difference is that we make the request to the same URL endpoint using the same HTTP method (`POST`). Instead of using HTTP verbs (`GET`, `POST`, `PUT`, `DELETE`) to express our action, we simply state our intended action in the body of the request. In the case of eAPI, we will specify a `method` key with a `runCmds` value for our intention.

For the following examples, I am using a physical Arista switch running EOS 4.16.

The eAPI preparation

The eAPI agent on the Arista device is disabled by default, so we will need to enable it on the device before we can use it:

```
arista1(config)#management api http-commands
arista1(config-mgmt-api-http-cmds)#no shut
arista1(config-mgmt-api-http-cmds)#protocol https port 443
arista1(config-mgmt-api-http-cmds)#no protocol http
arista1(config-mgmt-api-http-cmds)#vrf management
```

As you can see, we have turned off the HTTP server and are using HTTPS as the sole transport instead. Starting from a few EOS versions ago, the management interfaces, by default, reside in a VRF called **management**. In my topology, I am accessing the device via the management interface; therefore, I have specified the VRF for eAPI management. You can check that API management state via the "show management api http-commands" command:

```
arista1#sh management api http-commands
Enabled: Yes
HTTPS server: running, set to use port 443
HTTP server: shutdown, set to use port 80
Local HTTP server: shutdown, no authentication, set to use port 8080
Unix Socket server: shutdown, no authentication
VRF: management
Hits: 64
Last hit: 33 seconds ago
Bytes in: 8250
Bytes out: 29862
Requests: 23
Commands: 42
Duration: 7.086 seconds
SSL Profile: none
QoS DSCP: 0
 User Requests Bytes in Bytes out Last hit
 ----------- -------------- -------------- ---------------- --------------
  admin 23 8250 29862 33 seconds ago

URLs
-----------------------------------------
Management1 : https://192.168.199.158:443

arista1#
```

APIs and Intent-Driven Networking

After enabling the agent, you will be able to access the exploration page for eAPI by going to the device's IP address. If you have changed the default port for access, just append it at the end. The authentication is tied into the method of authentication on the switch. We will use the username and password configured locally on the device. By default, a self-signed certificate will be used:

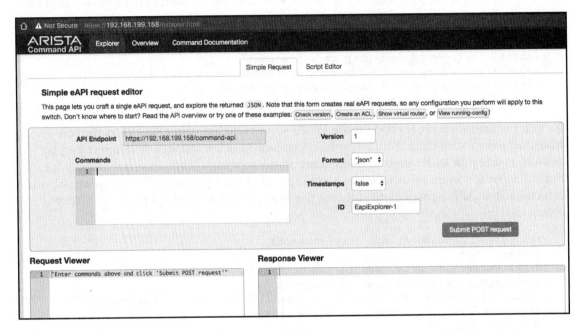

Arista EOS explorer

You will be taken to an explorer page where you can type in the CLI command and get a nice output for the body of your request. For example, if I want to see how to make a request body for `show version`, this is the output I will see from the explorer:

Chapter 3

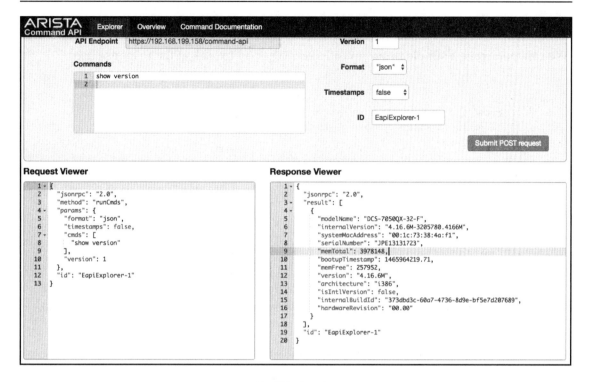

Arista EOS explorer viewer

The overview link will take you to the sample use and background information while the command documentation will serve as reference points for the show commands. Each of the command references will contain the returned value field name, type, and a brief description. The online reference scripts from Arista use jsonrpclib (https://github.com/joshmarshall/jsonrpclib/), which is what we will use. However, as of the time of writing this book, it has a dependency of Python 2.6+ and has not yet ported to Python 3; therefore, we will use Python 2.7 for these examples.

 By the time you read this book, there might be an updated status. Please read the GitHub pull request (https://github.com/joshmarshall/jsonrpclib/issues/38) and the GitHub README (https://github.com/joshmarshall/jsonrpclib/) for the latest status.

Installation is straightforward using `easy_install` or `pip`:

```
$ sudo easy_install jsonrpclib
$ sudo pip install jsonrpclib
```

[101]

eAPI examples

We can then write a simple program called `eapi_1.py` to look at the response text:

```
#!/usr/bin/python2

from __future__ import print_function
from jsonrpclib import Server
import ssl

ssl._create_default_https_context = ssl._create_unverified_context

switch = Server("https://admin:arista@192.168.199.158/command-api")

response = switch.runCmds( 1, [ "show version" ] )
print('Serial Number: ' + response[0]['serialNumber'])
```

> Note that, since this is Python 2, in the script, I used the `from __future__ import print_function` to make future migration easier. The `ssl`-related lines are for Python version > 2.7.9. For more information, please see https://www.python.org/dev/peps/pep-0476/.

This is the response I received from the previous `runCms()` method:

```
[{u'memTotal': 3978148, u'internalVersion': u'4.16.6M-
3205780.4166M', u'serialNumber': u'<omitted>', u'systemMacAddress':
u'<omitted>', u'bootupTimestamp': 1465964219.71, u'memFree':
277832, u'version': u'4.16.6M', u'modelName': u'DCS-7050QX-32-F',
u'isIntlVersion': False, u'internalBuildId': u'373dbd3c-60a7-4736-
8d9e-bf5e7d207689', u'hardwareRevision': u'00.00', u'architecture':
u'i386'}]
```

As you can see, the result is a list containing one dictionary item. If we need to grab the serial number, we can simply reference the item number and the key:

```
print('Serial Number: ' + response[0]['serialNumber'])
```

The output will contain only the serial number:

```
$ python eapi_1.py
Serial Number: <omitted>
```

To be more familiar with the command reference, I recommend that you click on the **Command Documentation** link on the eAPI page, and compare your output with the output of **show version** in the documentation.

As noted earlier, unlike REST, the JSON-RPC client uses the same URL endpoint for calling the server resources. You can see from the previous example that the `runCmds()` method contains a list of commands. For the execution of configuration commands, you can follow the same framework, and configure the device via a list of commands.

Here is an example of configuration commands named `eapi_2.py`. In our example, we wrote a function that takes the switch object and the list of commands as attributes:

```python
#!/usr/bin/python2

from __future__ import print_function
from jsonrpclib import Server
import ssl, pprint

ssl._create_default_https_context = ssl._create_unverified_context

# Run Arista commands thru eAPI
def runAristaCommands(switch_object, list_of_commands):
    response = switch_object.runCmds(1, list_of_commands)
    return response


switch = Server("https://admin:arista@192.168.199.158/command-api")

commands = ["enable", "configure", "interface ethernet 1/3", "switchport acc ess vlan 100", "end", "write memory"]

response = runAristaCommands(switch, commands)
pprint.pprint(response)
```

Here is the output of the command's execution:

```
$ python2 eapi_2.py
[{}, {}, {}, {}, {}, {u'messages': [u'Copy completed successfully.']}]
```

Now, do a quick check on the `switch` to verify the command's execution:

```
arista1#sh run int eth 1/3
interface Ethernet1/3
    switchport access vlan 100
arista1#
```

Overall, eAPI is fairly straightforward and simple to use. Most programming languages have libraries similar to `jsonrpclib`, which abstracts away JSON-RPC internals. With a few commands, you can start integrating Arista EOS automation into your network.

The Arista Pyeapi library

The Python client Pyeapi (http://pyeapi.readthedocs.io/en/master/index.html) library is a native Python library wrapper around eAPI. It provides a set of bindings to configure Arista EOS nodes. Why do we need Pyeapi when we already have eAPI? Picking between Pyeapi versus eAPI is mostly a judgment call if you are in a Python environment.

However, if you are in a non-Python environment, eAPI is probably the way to go. From our examples, you can see that the only requirement of eAPI is a JSON-RPC capable client. Thus, it is compatible with most programming languages. When I first started out in the field, Perl was the dominant language for scripting and network automation. There are still many enterprises that rely on Perl scripts as their primary automation tool. If you're in a situation where the company has already invested a ton of resources and the code base is in another language than Python, eAPI with JSON-RPC would be a good bet.

However, for those of us who prefer to code in Python, a native `Python` library means a more natural feeling in writing our code. It certainly makes extending a Python program to support the EOS node easier. It also makes keeping up with the latest changes in Python easier. For example, we can use Python 3 with Pyeapi!

At the time of writing this book, Python 3 (3.4+) support is officially a work-in-progress, as stated in the documentation (http://pyeapi.readthedocs.io/en/master/requirements.html). Please check the documentation for more details.

Pyeapi installation

Installation is straightforward with pip:

```
$ sudo pip install pyeapi
$ sudo pip3 install pyeapi
```

Note that pip will also install the netaddr library as it is part of the stated requirements (http://pyeapi.readthedocs.io/en/master/requirements.html) for Pyeapi.

By default, the Pyeapi client will look for an INI style hidden (with a period in front) file called `eapi.conf` in your home directory. You can override this behavior by specifying the `eapi.conf` file path, but it is generally a good idea to separate your connection credential and lock it down from the script itself. You can check out the Arista Pyeapi documentation (http://pyeapi.readthedocs.io/en/master/configfile.html#configfile) for the fields contained in the file. Here is the file I am using in the lab:

```
cat ~/.eapi.conf
[connection:Arista1]
host: 192.168.199.158
username: admin
password: arista
transport: https
```

The first line, `[connection:Arista1]`, contains the name that we will use in our Pyeapi connection; the rest of the fields should be pretty self-explanatory. You can lock down the file to be read-only for the user using this file:

```
$ chmod 400 ~/.eapi.conf
$ ls -l ~/.eapi.conf
-r-------- 1 echou echou 94 Jan 27 18:15 /home/echou/.eapi.conf
```

Pyeapi examples

Now, we are ready to take a look around the usage. Let's start by connecting to the EOS node by creating an object in the interactive Python shell:

```
Python 3.5.2 (default, Nov 17 2016, 17:05:23)
[GCC 5.4.0 20160609] on linux
Type "help", "copyright", "credits" or "license" for more information.
>>> import pyeapi
>>> arista1 = pyeapi.connect_to('Arista1')
```

We can execute show commands to the node and receive the output:

```
>>> import pprint
>>> pprint.pprint(arista1.enable('show hostname'))
[{'command': 'show hostname',
  'encoding': 'json',
  'result': {'fqdn': 'arista1', 'hostname': 'arista1'}}]
```

The configuration field can be either a single command or a list of commands using the `config()` method:

```
>>> arista1.config('hostname arista1-new')
[{}]
```

APIs and Intent-Driven Networking

```
>>> pprint.pprint(arista1.enable('show hostname'))
[{'command': 'show hostname',
  'encoding': 'json',
  'result': {'fqdn': 'arista1-new', 'hostname': 'arista1-new'}}]
>>> arista1.config(['interface ethernet 1/3', 'description my_link'])
[{}, {}]
```

Note that command abbreviation (`show run` versus `show running-config`) and some extensions will not work:

```
>>> pprint.pprint(arista1.enable('show run'))
Traceback (most recent call last):
...
 File "/usr/local/lib/python3.5/dist-packages/pyeapi/eapilib.py", line 396, in send
 raise CommandError(code, msg, command_error=err, output=out)
pyeapi.eapilib.CommandError: Error [1002]: CLI command 2 of 2 'show run' failed: invalid command [incomplete token (at token 1: 'run')]
>>>
>>> pprint.pprint(arista1.enable('show running-config interface ethernet 1/3'))
Traceback (most recent call last):
...
pyeapi.eapilib.CommandError: Error [1002]: CLI command 2 of 2 'show running-config interface ethernet 1/3' failed: invalid command [incomplete token (at token 2: 'interface')]
```

However, you can always catch the results and get the desired value:

```
>>> result = arista1.enable('show running-config')
>>> pprint.pprint(result[0]['result']['cmds']['interface Ethernet1/3'])
{'cmds': {'description my_link': None, 'switchport access vlan 100': None},
 'comments': []}
```

So far, we have been doing what we have been doing with eAPI for show and configuration commands. Pyeapi offers various APIs to make life easier. In the following example, we will connect to the node, call the VLAN API, and start to operate on the VLAN parameters of the device. Let's take a look:

```
>>> import pyeapi
>>> node = pyeapi.connect_to('Arista1')
>>> vlans = node.api('vlans')
>>> type(vlans)
<class 'pyeapi.api.vlans.Vlans'>
>>> dir(vlans)
[...'command_builder', 'config', 'configure', 'configure_interface',
'configure_vlan', 'create', 'default', 'delete', 'error', 'get',
'get_block', 'getall', 'items', 'keys', 'node', 'remove_trunk_group',
```

```
'set_name', 'set_state', 'set_trunk_groups', 'values']
>>> vlans.getall()
{'1': {'vlan_id': '1', 'trunk_groups': [], 'state': 'active', 'name':
'default'}}
>>> vlans.get(1)
{'vlan_id': 1, 'trunk_groups': [], 'state': 'active', 'name': 'default'}
>>> vlans.create(10)
True
>>> vlans.getall()
{'1': {'vlan_id': '1', 'trunk_groups': [], 'state': 'active', 'name':
'default'}, '10': {'vlan_id': '10', 'trunk_groups': [], 'state': 'active',
'name': 'VLAN0010'}}
>>> vlans.set_name(10, 'my_vlan_10')
True
```

Let's verify that VLAN 10 was created on the device:

```
arista1#sh vlan
VLAN Name Status Ports
----- -------------------------------- --------- -------------------------------
-----
1 default active
10 my_vlan_10 active
```

As you can see, the Python native API on the EOS object is really where Pyeapi excels beyond eAPI. It abstracts the lower-level attributes into the device object and makes the code cleaner and easier to read.

For a full list of ever increasing Pyeapi APIs, check the official documentation (http://pyeapi.readthedocs.io/en/master/api_modules/_list_of_modules.html).

To round up this chapter, let's assume that we repeat the previous steps enough times that we would like to write another Python class to save us some work.
The pyeapi_1.py script is shown as follows:

```
#!/usr/bin/env python3

import pyeapi

class my_switch():

    def __init__(self, config_file_location, device):
        # loads the config file
        pyeapi.client.load_config(config_file_location)
        self.node = pyeapi.connect_to(device)
```

APIs and Intent-Driven Networking

```
            self.hostname = self.node.enable('show hostname')[0]
    ['result']['host name']
            self.running_config = self.node.enable('show running-
config')

        def create_vlan(self, vlan_number, vlan_name):
            vlans = self.node.api('vlans')
            vlans.create(vlan_number)
            vlans.set_name(vlan_number, vlan_name)
```

As you can see from the script, we automatically connect to the node and set the hostname and `running_config` upon connection. We also create a method to the class that creates VLAN by using the `VLAN` API. Let's try out the script in an interactive shell:

```
Python 3.5.2 (default, Nov 17 2016, 17:05:23)
[GCC 5.4.0 20160609] on linux
Type "help", "copyright", "credits" or "license" for more information.
>>> import pyeapi_1
>>> s1 = pyeapi_1.my_switch('/tmp/.eapi.conf', 'Arista1')
>>> s1.hostname
'arista1'
>>> s1.running_config
[{'encoding': 'json', 'result': {'cmds': {'interface Ethernet27': {'cmds':
{}, 'comments': []}, 'ip routing': None, 'interface face Ethernet29':
{'cmds': {}, 'comments': []}, 'interface Ethernet26': {'cmds': {},
'comments': []}, 'interface Ethernet24/4': h.':
<omitted>
'interface Ethernet3/1': {'cmds': {}, 'comments': []}}, 'comments': [],
'header': ['! device: arista1 (DCS-7050QX-32, EOS-4.16.6M)n!n']},
'command': 'show running-config'}]
>>> s1.create_vlan(11, 'my_vlan_11')
>>> s1.node.api('vlans').getall()
{'11': {'name': 'my_vlan_11', 'vlan_id': '11', 'trunk_groups': [], 'state':
'active'}, '10': {'name': 'my_vlan_10', 'vlan_id': '10', 'trunk_groups':
[], 'state': 'active'}, '1': {'name': 'default', 'vlan_id': '1',
'trunk_groups': [], 'state': 'active'}}
>>>
```

Vendor-neutral libraries

There are several excellent efforts of vendor-neutral libraries such as Netmiko (`https://github.com/ktbyers/netmiko`) and NAPALM (`https://github.com/napalm-automation/napalm`). Because these libraries do not come natively from the device vendor, they are sometimes a step slower to support the latest platform or features. However, because the libraries are vendor-neutral, if you do not like vendor lock-in for your tools, then these libraries are a good choice. Another benefit of using these libraries is the fact that they are normally open source, so you can contribute back upstream for new features and bug fixes.

On the other hand, because these libraries are community supported, they are not necessarily the ideal fit if you need to rely on somebody else to fix bugs or implement new features. If you have a relatively small team that still needs to comply with certain service-level assurances for your tools, you might be better off using a vendor-backed library.

Summary

In this chapter, we looked at various ways to communicate and manage network devices from Cisco, Juniper, and Arista. We looked at both direct communication with the likes of NETCONF and REST, as well as using vendor-provided libraries such as PyEZ and Pyeapi. These are different layers of abstractions, meant to provide a way to programmatically manage your network devices without human intervention.

In `Chapter 4`, *The Python Automation Framework – Ansible Basics*, we will take a look at a higher level of vendor-neutral abstraction framework called **Ansible**. Ansible is an open source, general purpose automation tool written in Python. It can be used to automate servers, network devices, load balancers, and much more. Of course, for our purpose, we will focus on using this automation framework for network devices.

4
The Python Automation Framework – Ansible Basics

The previous two chapters incrementally introduced different ways to interact with network devices. In Chapter 2, *Low-Level Network Device Interactions*, we discussed Pexpect and Paramiko libraries that manage an interactive session to control the interactions. In Chapter 3, *APIs and Intent-Driven Networking*, we started to think of our network in terms of API and intent. We looked at various APIs that contain a well-defined command structure and provide a structured way of getting feedback from the device. As we moved from Chapter 2, *Low-Level Network Device Interactions*, to Chapter 3, *APIs and Intent-Driven Networking*, we began to think about our intent for the network and gradually expressed our network in terms of code.

Let's expand upon the idea of translating our intention into network requirements. If you have worked on network designs, chances are the most challenging part of the process is not the different pieces of network equipment, but rather qualifying and translating business requirements into the actual network design. Your network design needs to solve business problems. For example, you might be working within a larger infrastructure team that needs to accommodate a thriving online e-commerce site that experiences slow site response times during peak hours. How do you determine if the network is the problem? If the slow response on the website was indeed due to network congestion, which part of the network should you upgrade? Can the rest of the system take advantage of the bigger speed and feed? The following diagram is an illustration of a simple process of the steps that we might go through when trying to translate our business requirements into a network design:

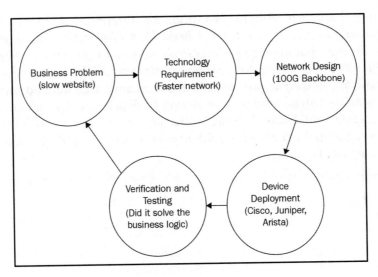

Business logic to network deployment

In my opinion, network automation is not just about faster configuration. It should also be about solving business problems, and accurately and reliably translating our intention into device behavior. These are the goals that we should keep in mind as we march on the network automation journey. In this chapter, we will start to look at a Python-based framework called **Ansible** that allows us to declare our intention for the network and abstract even more from the API and CLI.

A more declarative framework

You woke up one morning in a cold sweat from a nightmare you had about a potential network security breach. You realized that your network contains valuable digital assets that should be protected. You have been doing your job as a network administrator, so it is pretty secure, but you want to put more security measures around your network devices just to be sure.

To start with, you break the objective down into two actionable items:

- Upgrading the devices to the latest version of the software, which requires:
    1. Uploading the image to the device.
    2. Instructing the device to boot from the new image.
    3. Proceeding to reboot the device.
    4. Verifying that the device is running with the new software image.
- Configuring the appropriate access control list on the networking devices, which includes the following:
    1. Constructing the access list on the device.
    2. Configuring the access list on the interface, which in most cases is under the interface configuration section so that it can be applied to the interfaces.

Being an automation-focused network engineer, you want to write scripts to reliably configure the devices and receive feedback from the operations. You begin to research the necessary commands and APIs for each of the steps, validate them in the lab, and finally deploy them in production. Having done a fair amount of work for OS upgrade and ACL deployment, you hope the scripts are transferable to the next generation of devices. Wouldn't it be nice if there was a tool that could shorten this design-develop-deployment cycle?

In this chapter and in Chapter 5, *The Python Automation Framework – Beyond Basics*, we will work with an open source automation tool called **Ansible**. It is a framework that can simplify the process of going from business logic to network commands. It can configure systems, deploy software, and orchestrate a combination of tasks. Ansible is written in Python and has emerged as one of the leading automation tools supported by network equipment vendors.

In this chapter, we will take a look at the following topics:

- A quick Ansible example
- The advantages of Ansible

- The Ansible architecture
- Ansible Cisco modules and examples
- Ansible Juniper modules and examples
- Ansible Arista modules and examples

At the time of writing this book, Ansible release 2.5 is compatible with Python 2.6 and 2.7, with Python 3 support recently coming out of the technical review. Just like Python, many of the useful features of Ansible come from the community-driven extension modules. Even with Ansible core module supportability with Python 3, many of the extension modules and production deployments are still in Python 2 mode. It will take some time to bring all the extension modules up from Python 2 to Python 3. Due to this reason, for the rest of this book, we will use Python 2.7 with Ansible 2.2.

Why Ansible 2.2? Ansible 2.5, released in March 2018, offers many new network module features with a new connection method, syntax, and best practices. Given its relatively new features, most of the production deployment is still pre-2.5 release. However, in this chapter, you will also find sections dedicated to Ansible 2.5 examples for those who want to take advantage of the new syntax and features.

For the latest information on Ansible Python 3 support, check out `http://docs.ansible.com/ansible/python_3_support.html`.

As one can tell from the previous chapters, I am a believer in learning by examples. Just like the underlying Python code for Ansible, the syntax for Ansible constructs are easy enough to understand, even if you have not worked with Ansible before. If you have some experience with YAML or Jinja2, you will quickly draw the correlation between the syntax and the intended procedure. Let's take a look at an example first.

A quick Ansible example

As with other automation tools, Ansible started out by managing servers before expanding its ability to manage networking equipment. For the most part, the modules and what Ansible refers to as the playbook are similar between server modules and network modules with subtle differences. In this chapter, we will look at a server task example first and draw comparisons later on with network modules.

The control node installation

First, let's clarify the terminology we will use in the context of Ansible. We will refer to the virtual machine with Ansible installed as the control machine, and the machines being managed as the target machines or managed nodes. Ansible can be installed on most of the Unix systems, with the only dependency of Python 2.6 or 2.7. Currently, the Windows operating system is not officially supported as the control machine. Windows hosts can still be managed by Ansible, as they are just not supported as the control machine.

As Windows 10 starts to adopt the Windows Subsystem for Linux, Ansible might soon be ready to run on Windows as well. For more information, please check the Ansible documentation for Windows (https://docs.ansible.com/ansible/2.4/intro_windows.html).

On the managed node requirements, you may notice some documentation mentioning that Python 2.4 or later is a requirement. This is true for managing target nodes with operating systems such as Linux, but obviously not all network equipment supports Python. We will see how this requirement is bypassed for networking modules by local execution on the control node.

For Windows, Ansible modules are implemented in PowerShell. Windows modules in the core and extra repository live in a Windows/subdirectory if you would like to take a look.

We will be installing Ansible on our Ubuntu virtual machine. For instructions on installation on other operating systems, check out the installation documentation (http://docs.ansible.com/ansible/intro_installation.html). In the following code block, you will see the steps for installing the software packages:

```
$ sudo apt-get install software-properties-common
$ sudo apt-add-repository ppa:ansible/ansible
$ sudo apt-get update
$ sudo apt-get install ansible
```

We can also use `pip` to install Ansible: `pip install ansible`. My personal preference is to use the operating system's package management system, such as Apt on Ubuntu.

We can now do a quick verification as follows:

```
$ ansible --version
ansible 2.6.1
```

The Python Automation Framework – Ansible Basics

```
config file = /etc/ansible/ansible.cfg
```

Now, let's see how we can run different versions of Ansible on the same control node. This is a useful feature to adopt if you'd like to try out the latest development features without permanent installation. We can also use this method if we intend on running Ansible on a control node for which we do not have root permissions.

> As we saw from the output, at the time of writing this book, the latest release is 2.6.1. Feel free to use this version, but given the relatively new release, we will focus on Ansible version 2.2 in this book.

Running different versions of Ansible from source

You can run Ansible from a source code checkout (we will look at Git as a version control mechanism in Chapter 11, *Working with Git*):

```
$ git clone https://github.com/ansible/ansible.git --recursive
$ cd ansible/
$ source ./hacking/env-setup
...
Setting up Ansible to run out of checkout...
$ ansible --version
ansible 2.7.0.dev0 (devel cde3a03b32) last updated 2018/07/11 08:39:39 (GMT -700)
  config file = /etc/ansible/ansible.cfg
...
```

To run different versions, we can simply use `git checkout` for the different branch or tag and perform the environment setup again:

```
$ git branch -a
$ git tag --list
$ git checkout v2.5.6
...
HEAD is now at 0c985fe... New release v2.5.6
$ source ./hacking/env-setup
$ ansible --version
ansible 2.5.6 (detached HEAD 0c985fee8a) last updated 2018/07/11 08:48:20 (GMT -700)
  config file = /etc/ansible/ansible.cfg
```

[116]

If the Git commands seem a bit strange to you, we will cover Git in more detail in `Chapter 11`, *Working with Git*.

Once we are at the version you need, such as Ansible 2.2, we can run the update for the core modules for that version:

```
$ ansible --version
ansible 2.2.3.0 (detached HEAD f5be18f409) last updated 2018/07/14 07:40:09 (GMT -700)
...
$ git submodule update --init --recursive
Submodule 'lib/ansible/modules/core' (https://github.com/ansible/ansible-modules-core) registered for path 'lib/ansible/modules/core'
```

Let's take a look at the lab topology we will use in this chapter and `Chapter 5`, *The Python Automation Framework – Beyond Basics*.

Lab setup

In this chapter and in `Chapter 5`, *The Python Automation Framework – Beyond Basics*, our lab will have an Ubuntu 16.04 control node machine with Ansible installed. This control machine will have reachability for the management network for our VIRL devices, which consist of IOSv and NX-OSv devices. We will also have a separate Ubuntu VM for our playbook example when the target machine is a host:

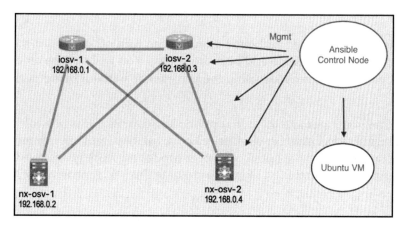

Lab topology

Now, we are ready to see our first Ansible playbook example.

Your first Ansible playbook

Our first playbook will be used between the control node and a remote Ubuntu host. We will take the following steps:

1. Make sure the control node can use key-based authorization.
2. Create an inventory file.
3. Create a playbook.
4. Execute and test it.

The public key authorization

The first thing to do is copy your SSH public key from your control machine to the target machine. A full public key infrastructure tutorial is outside the scope of this book, but here is a quick walkthrough on the control node:

```
$ ssh-keygen -t rsa <<<< generates public-private key pair on the host
machine if you have not done so already
$ cat ~/.ssh/id_rsa.pub <<<< copy the content of the output and paste it to
the ~/.ssh/authorized_keys file on the target host
```

You can read more about PKI at https://en.wikipedia.org/wiki/Public_key_infrastructure.

Because we are using key-based authentication, we can turn off password-based authentication on the remote node and be more secure. You will now be able to ssh from the control node to the remote node using the private key without being prompted for a password.

Can you automate the initial public key copying? It is possible, but is highly dependent on your use case, regulation, and environment. It is comparable to the initial console setup for network gears to establish initial IP reachability. Do you automate this? Why or why not?

The inventory file

We do not need Ansible if we have no remote target to manage, right? Everything starts with the fact that we need to perform some task on a remote host. In Ansible, the way we specify the potential remote target is with an inventory file. We can have this inventory file as the `/etc/ansible/hosts` file or use the `-i` option to specify the file during playbook runtime. Personally, I prefer to have this file in the same directory where my playbook is and use the `-i` option.

 Technically, this file can be named anything you like as long as it is in a valid format. However, the convention is to name this file `hosts`. You can potentially save yourself and your colleagues some headaches in the future by following this convention.

The inventory file is a simple, plaintext INI-style (https://en.wikipedia.org/wiki/INI_file) file that states your target. By default, the target can either be a DNS FQDN or an IP address:

```
$ cat hosts
192.168.199.170
```

We can now use the command-line option to test Ansible and the `hosts` file:

```
$ ansible -i hosts 192.168.199.170 -m ping
192.168.199.170 | SUCCESS => {
  "changed": false,
  "ping": "pong"
}
```

 By default, Ansible assumes that the same user executing the playbook exists on the remote host. For example, I am executing the playbook as `echou` locally; the same user also exists on my remote host. If you want to execute as a different user, you can use the `-u` option when executing, that is, `-u REMOTE_USER`.

The previous line in the example reads in the host file as the inventory file and executes the `ping` module on the host called `192.168.199.170`. Ping (http://docs.ansible.com/ansible/ping_module.html) is a trivial test module that connects to the remote host, verifies a usable Python installation, and returns the output `pong` upon success.

 You may take a look at the ever-expanding module list (http://docs.ansible.com/ansible/list_of_all_modules.html) if you have any questions about the use of existing modules that were shipped with Ansible.

If you get a host key error, it is typically because the host key is not in the `known_hosts` file, and is typically under `~/.ssh/known_hosts`. You can either SSH to the host and answer `yes` when adding the host, or you can disable this by checking on `/etc/ansible/ansible.cfg` or `~/.ansible.cfg` with the following code:

```
[defaults]
host_key_checking = False
```

Now that we have validated the inventory file and Ansible package, we can make our first playbook.

Our first playbook

Playbooks are Ansible's blueprint to describe what you would like to do to the hosts using modules. This is where we will be spending the majority of our time as operators when working with Ansible. If you are building a tree house, the playbook will be your manual, the modules will be your tools, while the inventory will be the components that you will be working on when using the tools.

The playbook is designed to be human readable, and is in YAML format. We will look at the common syntax used in the Ansible architecture section. For now, our focus is to run an example playbook to get the look and feel of Ansible.

Originally, YAML was said to mean Yet Another Markup Language, but now, `http://yaml.org/` has repurposed the acronym to be YAML ain't markup language.

Let's look at this simple 6-line playbook, `df_playbook.yml`:

```
---
- hosts: 192.168.199.170

  tasks:
    - name: check disk usage
      shell: df > df_temp.txt
```

Chapter 4

In a playbook, there can be one or more plays. In this case, we have one play (lines two to six). In any play, we can have one or more tasks. In our example play, we have just one task (lines four to six). The name field specifies the purpose of the task in a human readable format and the shell module was used. The module takes one argument of df. The shell module reads in the command in the argument and executes it on the remote host. In this case, we execute the df command to check the disk usage and copy the output to a file named df_temp.txt.

We can execute the playbook via the following code:

```
$ ansible-playbook -i hosts df_playbook.yml
PLAY [192.168.199.170]
************************************************************

TASK [setup]
******************************************************************
ok: [192.168.199.170]

TASK [check disk usage] *********************************************
changed: [192.168.199.170]

PLAY RECAP
******************************************************************
192.168.199.170  : ok=2    changed=1    unreachable=0    failed=0
```

If you log into the managed host (192.168.199.170, for me), you will see that the df_temp.txt file contains the output of the df command. Neat, huh?

You may have noticed that there were actually two tasks executed in our output, even though we only specified one task in the playbook; the setup module is automatically added by default. It is executed by Ansible to gather information about the remote host, which can be used later on in the playbook. For example, one of the facts that the setup module gathers is the operating system. What is the purpose of gathering facts about the remote target? You can use this information as a conditional for additional tasks in the same playbook. For example, the playbook can contain additional tasks to install packages. It can do this specifically to use apt for Debian-based hosts and yum for Red Hat-based hosts, based on the operation system facts that were gathered in the setup module.

> If you are curious about the output of a setup module, you can find out what information Ansible gathers via $ ansible -i hosts <host> -m setup.

[121]

Underneath the hood, there are actually a few things that have happened for our simple task. The control node copies the Python module to the remote host, executes the module, copies the module output to a temporary file, then captures the output and deletes the temporary file. For now, we can probably safely ignore these underlying details until we need them.

It is important that we fully understand the simple process that we have just gone through because we will be referring back to these elements later in this chapter. I purposely chose a server example to be presented here, because this will make more sense as we dive into the networking modules when we need to deviate from them (remember that we mentioned the Python interpreter is most likely not on the network gear).

Congratulations on executing your first Ansible playbook! We will look more into the Ansible architecture, but for now let's take a look at why Ansible is a good fit for network management. Remember that Ansible modules are written in Python? That is one advantage for a Pythonic network engineer, right?

The advantages of Ansible

There are many infrastructure automation frameworks besides Ansible—namely Chef, Puppet, and SaltStack. Each framework offers its own unique features and models; there is no one right framework that fits all the organizations. In this section, I would like to list some of the advantages of Ansible over other frameworks and why I think this is a good tool for network automation.

I am listing the advantages of Ansible without comparing them to other frameworks. Other frameworks might adopt some of the same philosophy or certain aspects of Ansible, but rarely do they contain all of the features that I will be mentioning. I believe it is the combination of all the following features and philosophy that makes Ansible ideal for network automation.

Agentless

Unlike some of its peers, Ansible does not require a strict master-client model. No software or agent needs to be installed on the client that communicates back to the server. Outside of the Python interpreter, which many platforms have by default, there is no additional software needed.

For network automation modules, instead of relying on remote host agents, Ansible uses SSH or API calls to push the required changes to the remote host. This further reduces the need for the Python interpreter. This is huge for network device management, as network vendors are typically reluctant to put third-party software on their platforms. SSH, on the other hand, already exists on the network equipment. This mentality has changed a bit in the last few years, but overall SSH is the common denominator for all network equipment while configuration management agent support is not. As you will remember from Chapter 2, *Low-Level Network Device Interactions*, newer network devices also provide an API layer, which can also be leveraged by Ansible.

Because there is no agent on the remote host, Ansible uses a push model to push the changes to the device, as opposed to the pull model where the agent pulls the information from the master server. The push model, in my opinion, is more deterministic as everything originates from the control machine. In a pull model, the timing of the `pull` might vary from client to client, and therefore results in change timing variance.

Again, the importance of being agentless cannot be stressed enough when it comes to working with the existing network equipment. This is usually one of the major reasons network operators and vendors embrace Ansible.

Idempotent

According to Wikipedia, idempotence is the property of certain operations in mathematics and computer science that can be applied multiple times without changing the result beyond the initial application (`https://en.wikipedia.org/wiki/Idempotence`). In more common terms, it means that running the same procedure over and over again does not change the system after the first time. Ansible aims to be idempotent, which is good for network operations that require a certain order of operations.

The advantage of idempotence is best compared to the Pexpect and Paramiko scripts that we have written. Remember that these scripts were written to push out commands as if an engineer was sitting at the terminal. If you were to execute the script 10 times, the script will make changes 10 times. If we write the same task via the Ansible playbook, the existing device configuration will be checked first, and the playbook will only execute if the changes do not exist. If we execute the playbook 10 times, the change will only be applied during the first run, with the next 9 runs suppressing the configuration change.

Being idempotent means we can repeatedly execute the playbook without worrying that there will be unnecessary changes made. This is important as we need to automatically check for state consistency without any extra overhead.

Simple and extensible

Ansible is written in Python and uses YAML for the playbook language, both of which are considered relatively easy to learn. Remember the Cisco IOS syntax? This is a domain-specific language that is only applicable when you are managing Cisco IOS devices or other similarly structured equipment; it is not a general purpose language beyond its limited scope. Luckily, unlike some other automation tools, there is no extra domain-specific language or DSL to learn for Ansible because YAML and Python are both widely used as general purpose languages.

As you can see from the previous example, even if you have not seen YAML before, it is easy to accurately guess what the playbook is trying to do. Ansible also uses Jinja2 as a template engine, which is a common tool used by Python web frameworks such as Django and Flask, so the knowledge is transferable.

I cannot stress enough the extensibility of Ansible. As illustrated by the preceding example, Ansible starts out with automating server (primarily Linux) workloads in mind. It then branches out to manage Windows machines with PowerShell. As more and more people in the industry started to adapt Ansible, the network became a topic that started to get more attention. The right people and team were hired at Ansible, network professionals started to get involved, and customers started to demand vendors for support. Starting with Ansible 2.0, network automation has become a first-class citizen alongside server management. The ecosystem is alive and well, with continuous improvement in each of the releases.

Just like the Python community, the Ansible community is friendly, and the attitude is inclusive of new members and ideas. I have first-hand experience of being a noob and trying to make sense of contribution procedures and wishing to write modules to be merged upstream. I can testify to the fact that I felt welcomed and respected for my opinions at all times.

The simplicity and extensibility really speak well for future proofing. The technology world is evolving fast, and we are constantly trying to adapt to it. Wouldn't it be great to learn a technology once and continue to use it, regardless of the latest trend? Obviously, nobody has a crystal ball to accurately predict the future, but Ansible's track record speaks well for future technology adaptation.

Network vendor support

Let's face it, we don't live in a vacuum. There is a running joke in the industry that the OSI layer should include a layer 8 (money) and 9 (politics). Every day, we need to work with network equipment made by various vendors.

Take API integration as an example. We saw the difference between the Pexpect and API approach in previous chapters. API clearly has an upper hand in terms of network automation. However, the API interface does not come cheap. Each vendor needs to invest time, money, and engineering resources to make the integration happen. The willingness for the vendor to support a technology matters greatly in our world. Luckily, all the major vendors support Ansible, as clearly indicated by the ever increasingly available network modules (http://docs.ansible.com/ansible/list_of_network_modules.html).

Why do vendors support Ansible more than other automation tools? Being agentless certainly helps, since having SSH as the only dependency greatly lowers the bar of entry. Engineers who have been on the vendor side know that the feature request process is usually months long and many hurdles have to be jumped through. Any time a new feature is added, it means more time spent on regression testing, compatibility checking, integration reviews, and many more. Lowering the bar of entry is usually the first step in getting vendor support.

The fact that Ansible is based on Python, a language liked by many networking professionals, is another great propeller for vendor support. For vendors such as Juniper and Arista who already made investments in PyEZ and Pyeapi, they can easily leverage the existing Python modules and quickly integrate their features into Ansible. As you will see in Chapter 5, *The Python Automation Framework – Beyond Basics*, we can use our existing Python knowledge to easily write our own modules.

Ansible already had a large number of community-driven modules before it focused on networking. The contribution process is somewhat baked and established, or as baked as an open source project can be. The core Ansible team is familiar with working with the community for submission and contribution.

Another reason for the increased network vendor support also has to do with Ansible's ability to give vendors the ability to express their own strength in the module context. We will see in the coming section that, besides SSH, the Ansible module can also be executed locally and communicate with these devices by using API. This ensures that vendors can express their latest and greatest features as soon as they make them available through the API. In terms of network professionals, this means that you can use the cutting-edge features to select the vendors when you are using Ansible as an automation platform.

We have spent a relatively large portion of space discussing vendor support because I feel that this is often an overlooked part in the Ansible story. Having vendors willing to put their weight behind the tool means you, the network engineer, can sleep at night knowing that the next big thing in networking will have a high chance of Ansible support, and you are not locked into your current vendor as your network needs to grow.

The Ansible architecture

The Ansible architecture consists of playbooks, plays, and tasks. Take a look at `df_playbook.yml` that we used previously:

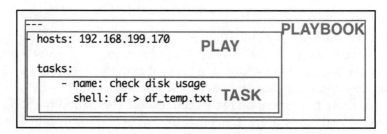

Ansible playbook

The whole file is called a playbook, which contains one or more plays. Each play can consist of one or more tasks. In our simple example, we only have one play, which contains a single task. In this section, we will take a look at the following:

- **YAML**: This format is extensively used in Ansible to express playbooks and variables.
- **Inventory**: The inventory is where you can specify and group hosts in your infrastructure. You can also optionally specify host and group variables in the inventory file.
- **Variables**: Each of the network devices is different. It has a different hostname, IP, neighbor relations, and so on. Variables allow for a standard set of plays while still accommodating these differences.
- **Templates**: Templates are nothing new in networking. In fact, you are probably using one without thinking of it as a template. What do we typically do when we need to provision a new device or replace an RMA (return merchandise authorization)? We copy the old configuration over and replace the differences such as the hostname and the loopback IP addresses. Ansible standardizes the template formatting with Jinja2, which we will dive deeper into later on.

In `Chapter 5`, *The Python Automation Framework – Beyond Basics*, we will cover some more advanced topics such as conditionals, loops, blocks, handlers, playbook roles, and how they can be included with network management.

YAML

YAML is the syntax used for Ansible playbooks and some other files. The official YAML documentation contains the full specifications of the syntax. Here is a compact version as it pertains to the most common usage for Ansible:

- A YAML file starts with three dashes (---)
- Whitespace indentation is used to denote structures when they are lined up, just like Python
- Comments begin with the hash (#) sign
- List members are denoted by a leading hyphen (-), with one member per line
- Lists can also be denoted via square brackets ([ ]), with elements separated by a comma (,)
- Dictionaries are denoted by key: value pairs, with a colon for separation
- Dictionaries can be denoted by curly braces, with elements separated by a comma (,)
- Strings can be unquoted, but can also be enclosed in double or single quotes

As you can see, YAML maps well into JSON and Python datatypes. If I were to rewrite `df_playbook.yml` into `df_playbook.json`, this is what it would look like:

```
[
  {
    "hosts": "192.168.199.170",
    "tasks": [
      "name": "check disk usage",
      "shell": "df > df_temp.txt"
    ]
  }
]
```

This is obviously not a valid playbook, but serves as an aid in helping to understand the YAML formats while using the JSON format as a comparison. Most of the time, comments (#), lists (-), and dictionaries (key: value) are what you will see in a playbook.

Inventories

By default, Ansible looks at the /etc/ansible/hosts file for hosts specified in your playbook. As mentioned previously, I find it more expressive to specify the host file via the -i option. This is what we have been doing up to this point. To expand on our previous example, we can write our inventory host file as follows:

```
[ubuntu]
192.168.199.170

[nexus]
192.168.199.148
192.168.199.149

[nexus:vars]
username=cisco
password=cisco

[nexus_by_name]
switch1 ansible_host=192.168.199.148
switch2 ansible_host=192.168.199.149
```

As you may have guessed, the square bracket headings specify group names, so later on in the playbook we can point to this group. For example, in cisco_1.yml and cisco_2.yml, I can act on all of the hosts specified under the nexus group to the group name of nexus:

```
---
- name: Configure SNMP Contact
  hosts: "nexus"
  gather_facts: false
  connection: local
<skip>
```

A host can exist in more than one group. The group can also be nested as children:

```
[cisco]
router1
router2

[arista]
switch1
switch2

[datacenter:children]
cisco
arista
```

[128]

In the previous example, the datacenter group includes both the `cisco` and `arista` members.

We will discuss variables in the next section. However, you can optionally specify variables belonging to the host and group in the inventory file as well. In our first inventory file example, [nexus:vars] specifies variables for the whole nexus group. The `ansible_host` variable declares variables for each of the hosts on the same line.

For more information on the inventory file, check out the official documentation (http://docs.ansible.com/ansible/intro_inventory.html).

Variables

We discussed variables a bit in the previous section. Because our managed nodes are not exactly alike, we need to accommodate the differences via variables. Variable names should be letters, numbers, and underscores, and should always start with a letter. Variables are commonly defined in three locations:

- The playbook
- The inventory file
- Separate files to be included in files and roles

Let's look at an example of defining variables in a playbook, `cisco_1.yml`:

```
---
- name: Configure SNMP Contact
  hosts: "nexus"
  gather_facts: false
  connection: local

  vars:
  cli:
    host: "{{ inventory_hostname }}"
    username: cisco
    password: cisco
    transport: cli

  tasks:
  - name: configure snmp contact
    nxos_snmp_contact:
      contact: TEST_1
      state: present
      provider: "{{ cli }}"
```

The Python Automation Framework – Ansible Basics

```
    register: output

  - name: show output
    debug:
      var: output
```

You can see the `cli` variable declared under the `vars` section, which is being used in the task of `nxos_snmp_contact`.

> For more information on the `nxso_snmp_contact` module, check out the online documentation (http://docs.ansible.com/ansible/nxos_snmp_contact_module.html).

To reference a variable, you can use the Jinja2 templating system convention of a double curly bracket. You don't need to put quotes around the curly bracket unless you are starting a value with it. I typically find it easier to remember and put a quote around the variable value regardless.

You may have also noticed the `{{ inventory_hostname }}` reference, which is not declared in the playbook. It is one of the default variables that Ansible provides for you automatically, and it is sometimes referred to as the magic variable.

> There are not many magic variables, and you can find the list in the documentation (http://docs.ansible.com/ansible/playbooks_variables.html#magic-variables-and-how-to-access-information-about-other-hosts).

We have declared variables in an inventory file in the previous section:

```
[nexus:vars]
username=cisco
password=cisco

[nexus_by_name]
switch1 ansible_host=192.168.199.148
switch2 ansible_host=192.168.199.149
```

To use the variables in the inventory file instead of declaring them in the playbook, let's add the group variables for `[nexus_by_name]` in the host file:

```
[nexus_by_name]
switch1 ansible_host=192.168.199.148
switch2 ansible_host=192.168.199.149

[nexus_by_name:vars]
```

```
username=cisco
password=cisco
```

Then, modify the playbook to match what we can see here in `cisco_2.yml`, to reference the variables:

```yaml
---
- name: Configure SNMP Contact
  hosts: "nexus_by_name"
  gather_facts: false
  connection: local

  vars:
    cli:
      host: "{{ ansible_host }}"
      username: "{{ username }}"
      password: "{{ password }}"
      transport: cli

  tasks:
    - name: configure snmp contact
      nxos_snmp_contact:
        contact: TEST_1
        state: present
        provider: "{{ cli }}"

      register: output

    - name: show output
      debug:
        var: output
```

Notice that in this example, we are referring to the `nexus_by_name` group in the inventory file, the `ansible_host` host variable, and the `username` and `password` group variables. This is a good way of hiding the username and password in a write-protected file and publish the playbook without the fear of exposing your sensitive data.

 To see more examples of variables, check out the Ansible documentation (http://docs.ansible.com/ansible/playbooks_variables.html).

To access complex variable data that's provided in a nested data structure, you can use two different notations. Noted in the `nxos_snmp_contact` task, we registered the output in a variable and displayed it using the debug module. You will see something like the following during playbook execution:

```
TASK [show output]
************************************************************
ok: [switch1] => {
 "output": {
  "changed": false,
    "end_state": {
       "contact": "TEST_1"
     },
    "existing": {
       "contact": "TEST_1"
     },
    "proposed": {
       "contact": "TEST_1"
     },
     "updates": []
    }
  }
}
```

In order to access the nested data, we can use the following notation, as specified in `cisco_3.yml`:

```
msg: '{{ output["end_state"]["contact"] }}'
msg: '{{ output.end_state.contact }}'
```

You will receive just the value indicated:

```
TASK [show output in output["end_state"]["contact"]]
***************************
ok: [switch1] => {
 "msg": "TEST_1"
}
ok: [switch2] => {
 "msg": "TEST_1"
}

TASK [show output in output.end_state.contact]
********************************
ok: [switch1] => {
 "msg": "TEST_1"
}
ok: [switch2] => {
 "msg": "TEST_1"
}
```

Lastly, we mentioned variables can also be stored in a separate file. To see how we can use variables in a role or included file, we should get a few more examples under our belt, because they are a bit complicated to start with. We will see more examples of roles in Chapter 5, *The Python Automation Framework – Beyond Basics*.

Templates with Jinja2

In the previous section, we used variables with the Jinja2 syntax of `{{ variable }}`. While you can do a lot of complex things in Jinja2, luckily, we only need some of the basic things to get started.

> Jinja2 (http://jinja.pocoo.org/) is a full-featured, powerful template engine that originated in the Python community. It is widely used in Python web frameworks such as Django and Flask.

For now, it is enough to just keep in mind that Ansible utilizes Jinja2 as the template engine. We will revisit the topics of Jinja2 filters, tests, and lookups as the situations call for them. You can find more information on the Ansible Jinja2 template here: http://docs.ansible.com/ansible/playbooks_templating.html.

Ansible networking modules

Ansible was originally made for managing nodes with full operating systems such as Linux and Windows before it was extended to support network equipment. You may have already noticed the subtle differences in playbooks that we have used so far for network devices, such as the lines of `gather_facts: false` and `connection: local`; we will take a closer look at the differences in the following sections.

Local connections and facts

Ansible modules are Python code that's executed on the remote host by default. Because of the fact that most network equipment does not expose Python directly, or they simply do not contain Python, we are almost always executing the playbook locally. This means that the playbook is interpreted locally first and commands or configurations are pushed out later on as needed.

Recall that the remote host facts were gathered via the setup module, which was added by default. Since we are executing the playbook locally, the setup module will gather the facts on the localhost instead of the remote host. This is certainly not needed, therefore when the connection is set to local, we can reduce this unnecessary step by setting the fact gathering to false.

Because network modules are executed locally, for those modules that offer a backup option, the files are backed up locally on the control node as well.

One of the most important changes in Ansible 2.5 was the introduction of different communication protocols (https://docs.ansible.com/ansible/latest/network/getting_started/network_differences.html#multiple-communication-protocols). The connection method now includes `network_cli`, `netconf`, `httpapi`, and `local`. If the network device uses CLI over SSH, you indicate the connection method as `network_cli` in one of the device variables. However, due to the fact that this is a relatively recent change, you might still see the connection stated as local in many of the existing playbooks.

Provider arguments

As we have seen from Chapter 2, *Low-Level Network Device Interactions*, and Chapter 3, *APIs and Intent-Driven Networking*, network equipment can be connected via both SSH or API, depending on the platform and software release. All core networking modules implement a `provider` argument, which is a collection of arguments used to define how to connect to the network device. Some modules only support `cli` while some support other values, for example, Arista EAPI and Cisco NXAPI. This is where Ansible's "let the vendor shine" philosophy is demonstrated. The module will have documentation on which transport method they support.

Starting with Ansible 2.5, the recommended way to specify the transport method is by using the `connection` variable. You will start to see the provider parameter being gradually phased out from future Ansible releases. Using the `ios_command` module as an example, https://docs.ansible.com/ansible/latest/modules/ios_command_module.html#ios-command-module, the provider parameter still works, but is being labeled as deprecated. We will see an example of this later in this chapter.

Some of the basic arguments supported by the `provider` transport are as follows:

- `host`: This defines the remote host
- `port`: This defines the port to connect to
- `username`: This is the username to be authenticated

- `password`: This is the password to be authenticated
- `transport`: This is the type of transport for the connection
- `authorize`: This enables privilege escalation for devices that require it
- `auth_pass`: This defines the privilege escalation password

As you can see, not all arguments need to be specified. For example, for our previous playbooks, our user is always at the admin privilege when logged in, therefore we do not need to specify the `authorize` or the `auth_pass` arguments.

These arguments are just variables, so they follow the same rules for variable precedence. For example, if I change `cisco_3.yml` to `cisco_4.yml` and observe the following precedence:

```
---
- name: Configure SNMP Contact
  hosts: "nexus_by_name"
  gather_facts: false
  connection: local

  vars:
    cli:
      host: "{{ ansible_host }}"
      username: "{{ username }}"
      password: "{{ password }}"
      transport: cli

  tasks:
    - name: configure snmp contact
      nxos_snmp_contact:
        contact: TEST_1
        state: present
        username: cisco123
        password: cisco123
        provider: "{{ cli }}"

      register: output

    - name: show output in output["end_state"]["contact"]
      debug:
        msg: '{{ output["end_state"]["contact"] }}'

    - name: show output in output.end_state.contact
      debug:
        msg: '{{ output.end_state.contact }}'
```

The username and password defined on the task level will override the username and password at the playbook level. I will receive the following error when trying to connect because the user does not exist on the device:

```
PLAY [Configure SNMP Contact] ****************************************

TASK [configure snmp contact] ****************************************
fatal: [switch2]: FAILED! => {"changed": false, "failed": true,
"msg": "failed to connect to 192.168.199.149:22"}
fatal: [switch1]: FAILED! => {"changed": false, "failed": true,
"msg": "failed to connect to 192.168.199.148:22"}
to retry, use: --limit
@/home/echou/Master_Python_Networking/Chapter4/cisco_4.retry

PLAY RECAP ***********************************************************
switch1 : ok=0 changed=0 unreachable=0 failed=1
switch2 : ok=0 changed=0 unreachable=0 failed=1
```

The Ansible Cisco example

Cisco's support in Ansible is categorized by the operating systems IOS, IOS-XR, and NX-OS. We have already seen a number of NX-OS examples, so in this section let's try to manage IOS-based devices.

Our host file will consist of two hosts, `R1` and `R2`:

```
[ios_devices]
R1 ansible_host=192.168.24.250
R2 ansible_host=192.168.24.251

[ios_devices:vars]
username=cisco
password=cisco
```

Our playbook, `cisco_5.yml`, will use the `ios_command` module to execute arbitrary show commands:

```
---
- name: IOS Show Commands
  hosts: "ios_devices"
  gather_facts: false
  connection: local
```

```
     vars:
       cli:
         host: "{{ ansible_host }}"
         username: "{{ username }}"
         password: "{{ password }}"
         transport: cli

     tasks:
       - name: ios show commands
         ios_command:
           commands:
             - show version | i IOS
             - show run | i hostname
           provider: "{{ cli }}"

         register: output

       - name: show output in output["end_state"]["contact"]
         debug:
           var: output
```

The result is what we would expect as the `show version` and `show run` output:

```
$ ansible-playbook -i ios_hosts cisco_5.yml

PLAY [IOS Show Commands]
***********************************************************

TASK [ios show commands]
***********************************************************
ok: [R1]
ok: [R2]

TASK [show output in output["end_state"]["contact"]]
***************************
ok: [R1] => {
 "output": {
 "changed": false,
 "stdout": [
 "Cisco IOS Software, 7200 Software (C7200-A3JK9S-M), Version
12.4(25g), RELEASE SOFTWARE (fc1)",
 "hostname R1"
 ],
 "stdout_lines": [
 [
 "Cisco IOS Software, 7200 Software (C7200-A3JK9S-M), Version
12.4(25g), RELEASE SOFTWARE (fc1)"
 ],
```

```
        [
         "hostname R1"
        ]
       ]
      }
    }
    ok: [R2] => {
      "output": {
       "changed": false,
       "stdout": [
        "Cisco IOS Software, 7200 Software (C7200-A3JK9S-M), Version
    12.4(25g), RELEASE SOFTWARE (fc1)",
        "hostname R2"
       ],
       "stdout_lines": [
        [
         "Cisco IOS Software, 7200 Software (C7200-A3JK9S-M), Version
    12.4(25g), RELEASE SOFTWARE (fc1)"
        ],
        [
         "hostname R2"
        ]
       ]
      }
    }

    PLAY RECAP
    **********************************************************************
    R1    : ok=2    changed=0    unreachable=0    failed=0
    R2    : ok=2    changed=0    unreachable=0    failed=0
```

I wanted to point out a few things illustrated by this example:

- The playbook between NXOS and IOS is largely identical
- The syntax `nxos_snmp_contact` and `ios_command` modules follow the same pattern, with the only difference being the argument for the modules
- The IOS version of the devices are pretty old with no understanding of API, but the modules still have the same look and feel

As you can see from the preceding example, once we have the basic syntax down for the playbooks, the subtle difference relies on the different modules for the task we would like to perform.

Ansible 2.5 connection example

We have briefly talked about the addition of network connection changes in Ansible playbooks, starting with version 2.5. Along with the changes, Ansible also released a network best practices document, https://docs.ansible.com/ansible/latest/network/user_guide/network_best_practices_2.5.html. Let's build an example based on the best practices guide. For our topology, we will reuse the topology in Chapter 2, *Low-Level Network Device Interactions*, with two IOSv devices. Since there are multiple files involved in this example, the files are grouped into a subdirectory named ansible_2-5_example.

Our inventory file is reduced to the group and the name of the hosts:

```
$ cat hosts
[ios-devices]
iosv-1
iosv-2
```

We have created a host_vars directory with two files. Each corresponds to the name specified in the inventory file:

```
$ ls -a host_vars/
. .. iosv-1 iosv-2
```

The variable file for the hosts contains what was previously included in the CLI variable. The additional variable of ansible_connection specifies network_cli as the transport:

```
$ cat host_vars/iosv-1
---
ansible_host: 172.16.1.20
ansible_user: cisco
ansible_ssh_pass: cisco
ansible_connection: network_cli
ansible_network_os: ios
ansbile_become: yes
ansible_become_method: enable
ansible_become_pass: cisco

$ cat host_vars/iosv-2
---
ansible_host: 172.16.1.21
ansible_user: cisco
ansible_ssh_pass: cisco
ansible_connection: network_cli
ansible_network_os: ios
ansbile_become: yes
ansible_become_method: enable
```

```
    ansible_become_pass: cisco
```

Our playbook will use the `ios_config` module with the `backup` option enabled. Notice the use of the `when` condition in this example so that if there are other hosts with a different operating system, this task will not be applied:

```
$ cat my_playbook.yml
---
- name: Chapter 4 Ansible 2.5 Best Practice Demonstration
  connection: network_cli
  gather_facts: false
  hosts: all
  tasks:
    - name: backup
      ios_config:
        backup: yes
      register: backup_ios_location
      when: ansible_network_os == 'ios'
```

When the playbook is run, a new backup folder will be created with the configuration backed up for each of the hosts:

```
$ ansible-playbook -i hosts my_playbook.yml

PLAY [Chapter 4 Ansible 2.5 Best Practice Demonstration]
***********************

TASK [backup]
***************************************************************
ok: [iosv-2]
ok: [iosv-1]

PLAY RECAP
****************************************************************
iosv-1 : ok=1  changed=0  unreachable=0  failed=0
iosv-2 : ok=1  changed=0  unreachable=0  failed=0

$ ls -l backup/
total 8
-rw-rw-r-- 1 echou echou 3996 Jul 11 19:01
iosv-1_config.2018-07-11@19:01:55
-rw-rw-r-- 1 echou echou 3996 Jul 11 19:01
iosv-2_config.2018-07-11@19:01:55

$ cat backup/iosv-1_config.2018-07-11@19\:01\:55
Building configuration...
```

```
Current configuration : 3927 bytes
!
! Last configuration change at 01:46:00 UTC Thu Jul 12 2018 by cisco
!
version 15.6
service timestamps debug datetime msec
service timestamps log datetime msec
...
```

This example illustrates the `network_connection` variable and the recommended structure based on network best practices. We will look at offloading variables into the `host_vars` directory and conditionals in Chapter 5, *The Python Automation Framework – Beyond Basics*. This structure can also be used for the Juniper and Arista examples in this chapter. For the different devices, we will just use different values for `network_connection`.

The Ansible Juniper example

The Ansible Juniper module requires the Juniper PyEZ package and NETCONF. If you have been following the API example in Chapter 3, *APIs and Intent-Driven Networking*, you are good to go. If not, refer back to that section for installation instructions as well as some test script to make sure PyEZ works. The Python package called `jxmlease` is also required:

```
$ sudo pip install jxmlease
```

In the host file, we will specify the device and connection variables:

```
[junos_devices]
J1 ansible_host=192.168.24.252

[junos_devices:vars]
username=juniper
password=juniper!
```

In our Juniper playbook, we will use the `junos_facts` module to gather basic facts for the device. This module is equivalent to the setup module and will come in handy if we need to take action depending on the returned value. Note the different value of transport and port in the example here:

```
---
- name: Get Juniper Device Facts
  hosts: "junos_devices"
  gather_facts: false
  connection: local
```

```yaml
    vars:
      netconf:
        host: "{{ ansible_host }}"
        username: "{{ username }}"
        password: "{{ password }}"
        port: 830
        transport: netconf

    tasks:
      - name: collect default set of facts
        junos_facts:
          provider: "{{ netconf }}"

        register: output

      - name: show output
        debug:
          var: output
```

When executed, you will receive this output from the `Juniper` device:

```
PLAY [Get Juniper Device Facts]
*************************************************

TASK [collect default set of facts]
*****************************************
ok: [J1]

TASK [show output]
***********************************************************
ok: [J1] => {
"output": {
"ansible_facts": {
"HOME": "/var/home/juniper",
"domain": "python",
"fqdn": "master.python",
"has_2RE": false,
"hostname": "master",
"ifd_style": "CLASSIC",
"model": "olive",
"personality": "UNKNOWN",
"serialnumber": "",
"switch_style": "NONE",
"vc_capable": false,
"version": "12.1R1.9",
"version_info": {
"build": 9,
"major": [
```

```
            12,
            1
        ],
        "minor": "1",
        "type": "R"
        }
    },
    "changed": false
    }
}

PLAY RECAP *********************************************************
J1   : ok=2    changed=0    unreachable=0    failed=0
```

The Ansible Arista example

The final playbook example we will look at will be the Arista command module. At this point, we are quite familiar with our playbook syntax and structure. The Arista device can be configured to use transport using `cli` or `eapi`, so, in this example, we will use `cli`.

This is the host file:

```
[eos_devices]
A1 ansible_host=192.168.199.158
```

The playbook is also similar to what we have seen previously:

```
---
- name: EOS Show Commands
  hosts: "eos_devices"
  gather_facts: false
  connection: local

  vars:
    cli:
      host: "{{ ansible_host }}"
      username: "arista"
      password: "arista"
      authorize: true
      transport: cli

  tasks:
    - name: eos show commands
      eos_command:
        commands:
```

```yaml
   - show version | i Arista
     provider: "{{ cli }}"
     register: output

   - name: show output
     debug:
       var: output
```

The output will show the standard output as we would expect from the command line:

```
PLAY [EOS Show Commands]
****************************************************************

TASK [eos show commands]
****************************************************************
ok: [A1]

TASK [show output]
****************************************************************
ok: [A1] => {
  "output": {
  "changed": false,
  "stdout": [
  "Arista DCS-7050QX-32-F"
  ],
  "stdout_lines": [
  [
  "Arista DCS-7050QX-32-F"
  ]
  ],
  "warnings": []
  }
}

PLAY RECAP
****************************************************************
A1 : ok=2    changed=0    unreachable=0    failed=0
```

Summary

In this chapter, we took a grand tour of the open source automation framework Ansible. Unlike Pexpect-based and API-driven network automation scripts, Ansible provides a higher layer of abstraction called the playbook to automate our network devices.

Ansible was originally constructed to manage servers and was later extended to network devices; therefore we took a look at a server example. Then, we compared and contrasted the differences when it came to network management playbooks. Later, we looked at the example playbooks for Cisco IOS, Juniper JUNOS, and Arista EOS devices. We also looked at the best practices recommended by Ansible if you are using Ansible version 2.5 and later.

In Chapter 5, *The Python Automation Framework – Beyond Basics*, we will leverage the knowledge we gained in this chapter and start to look at some of the more advanced features of Ansible.

5
The Python Automation Framework – Beyond Basics

In Chapter 1, *Review of TCP/IP Protocol Suite and Python*, we looked at some of the basic structures to get Ansible up and running. We worked with Ansible inventory files, variables, and playbooks. We also looked at some examples of using network modules for Cisco, Juniper, and Arista devices.

In this chapter, we will further build on the knowledge we have gained from the previous chapters and dive deeper into the more advanced topics of Ansible. Many books have been written about Ansible, and there is more to Ansible than we can cover in two chapters. The goal here is to introduce the majority of the features and functions of Ansible that I believe you will need as a network engineer and shorten the learning curve as much as possible.

It is important to point out that if you were not clear on some of the points made in Chapter 4, *The Python Automation Framework – Ansible Basics*, now is a good time to go back and review them as they are a prerequisite for this chapter.

In this chapter, we will look into the following topics:

- Ansible conditionals
- Ansible loops
- Templates
- Group and host variables
- The Ansible Vault
- Ansible roles
- Writing your own module

We have a lot of ground to cover, so let's get started!

Ansible conditionals

Ansible conditionals are similar to conditional statements in programming languages. In Chapter 1, *Review of TCP/IP Protocol Suite and Python*, we saw that Python uses conditional statements to only execute a section of the code using `if..then` or `while` statements. In Ansible, it uses conditional keywords to only run a task when the condition is met. In many cases, the execution of a play or task may depend on the value of a fact, variable, or the previous task result. For example, if you have a play to upgrading router images, you want to include a step to make sure the new router image is on the device before you move on to the next play of rebooting the router.

In this section, we will discuss the `when` clause, which is supported for all modules, as well as unique conditional states that are supported in Ansible networking command modules. Some of the conditions are as follows:

- Equal to (`eq`)
- Not equal to (`neq`)
- Greater than (`gt`)
- Greater than or equal to (`ge`)
- Less than (`lt`)
- Less than or equal to (`le`)
- Contains

The when clause

The `when` clause is useful when you need to check the output of a variable or a play execution result and act accordingly. We saw a quick example of the `when` clause in Chapter 4, *The Python Automation Framework – Ansible Basics*, when we looked at the Ansible 2.5 best practices structure. If you recall, the task only ran when the network operating system of the device was the Cisco IOS. Let's look at another example of its use in `chapter5_1.yml`:

```
---
- name: IOS Command Output
  hosts: "iosv-devices"
  gather_facts: false
  connection: local
  vars:
    cli:
      host: "{{ ansible_host }}"
```

```
          username: "{{ username }}"
          password: "{{ password }}"
          transport: cli
  tasks:
    - name: show hostname
      ios_command:
        commands:
          - show run | i hostname
        provider: "{{ cli }}"
      register: output
    - name: show output
      when: '"iosv-2" in "{{ output.stdout }}"'
      debug:
        msg: '{{ output }}'
```

We have seen all the elements in this playbook before in Chapter 4, *The Python Automation Framework – Ansible Basics*, up to the end of the first task. For the second task in the play, we are using the when clause to check if the output contains the iosv-2 keyword. If true, we will proceed to the task, which is using the debug module to display the output. When the playbook is run, we will see the following output:

```
<skip>
TASK [show output]
*************************************************************
skipping: [ios-r1]
ok: [ios-r2] => {
    "msg": {
        "changed": false,
        "stdout": [
            "hostname iosv-2"
        ],
        "stdout_lines": [
            [
                "hostname iosv-2"
            ]
        ],
        "warnings": []
    }
}
<skip>
```

We can see that the iosv-r1 device is skipped from the output because the clause did not pass. We can further expand this example in chapter5_2.yml to only apply certain configuration changes when the condition is met:

```
<skip>
tasks:
```

```
    - name: show hostname
      ios_command:
        commands:
          - show run | i hostname
        provider: "{{ cli }}"
      register: output
    - name: config example
      when: '"iosv-2" in "{{ output.stdout }}"'
      ios_config:
        lines:
          - logging buffered 30000
        provider: "{{ cli }}"
```

We can see the execution output here:

```
TASK [config example]
************************************************************
skipping: [ios-r1]
changed: [ios-r2]

PLAY RECAP
************************************************************
ios-r1 : ok=1 changed=0 unreachable=0 failed=0
ios-r2 : ok=2 changed=1 unreachable=0 failed=0
```

Again, note in the execution output that `ios-r2` was the only change applied while `ios-r1` was skipped. In this case, the logging buffer size was only changed on `ios-r2`.

The `when` clause is also very useful in situations when the setup or facts module is used – you can act based on some of the `facts` that were gathered initially. For example, the following statement will ensure that only the Ubuntu host with major release `16` will be acted upon by placing a conditional statement in the clause:

```
when: ansible_os_family == "Debian" and ansible_lsb.major_release|int >= 16
```

 For more conditionals, check out the Ansible conditionals documentation (http://docs.ansible.com/ansible/playbooks_conditionals.html).

Ansible network facts

Prior to 2.5, Ansible networking shipped with a number of network-specific fact modules. The network fact modules exist, but the naming and usage was different between vendors. Starting with version 2.5, Ansible started to standardize its network fact module usage. The Ansible network fact modules gather information from the system and store the results in facts prefixed with `ansible_net_`. The data collected by these modules is documented in the *return values* in the module documentation. This is a pretty big milestone for Ansible networking modules, as it does a lot of the heavy lifting for you to abstract the fact-gathering process by default.

Let's use the same structure we saw in Chapter 4, *The Python Automation Framework – Ansible Basics*, Ansible 2.5 best practices, but expand upon it to see how the `ios_facts` module was used to gather facts. As a review, our inventory file contains two iOS hosts with the host variables residing in the `host_vars` directory:

```
$ cat hosts
[ios-devices]
iosv-1
iosv-2

$ cat host_vars/iosv-1
---
ansible_host: 172.16.1.20
ansible_user: cisco
ansible_ssh_pass: cisco
ansible_connection: network_cli
ansible_network_os: ios
ansbile_become: yes
ansible_become_method: enable
ansible_become_pass: cisco
```

Our playbook will have three tasks. The first task will use the `ios_facts` module to gather facts for both of our network devices. The second task will display certain facts gathered and stored for each of the two devices. You will see that the facts we displayed were the default `ansible_net` facts, as opposed to a registered variable from the first task. The third task will display all the facts we collected for the `iosv-1` host:

```
$ cat my_playbook.yml
---
- name: Chapter 5 Ansible 2.5 network facts
  connection: network_cli
  gather_facts: false
  hosts: all
  tasks:
```

The Python Automation Framework – Beyond Basics

```
    - name: Gathering facts via ios_facts module
      ios_facts:
      when: ansible_network_os == 'ios'

    - name: Display certain facts
      debug:
        msg: "The hostname is {{ ansible_net_hostname }} running {{ ansible_net_version }}"

    - name: Display all facts for a host
      debug:
        var: hostvars['iosv-1']
```

When we run the playbook, you can see that the result for the first two tasks were what we would have expected:

```
$ ansible-playbook -i hosts my_playbook.yml

PLAY [Chapter 5 Ansible 2.5 network facts]
*************************************

TASK [Gathering facts via ios_facts module]
*************************************
ok: [iosv-2]
ok: [iosv-1]

TASK [Display certain facts]
*****************************************************
ok: [iosv-2] => {
    "msg": "The hostname is iosv-2 running 15.6(3)M2"
}
ok: [iosv-1] => {
    "msg": "The hostname is iosv-1 running 15.6(3)M2"
}
```

The third task will display all the network device facts gathered for iOS devices. There is a ton of information that has been gathered for iOS devices that can help with your networking automation needs:

```
TASK [Display all facts for a host]
*********************************************
ok: [iosv-1] => {
    "hostvars['iosv-1']": {
        "ansbile_become": true,
        "ansible_become_method": "enable",
        "ansible_become_pass": "cisco",
        "ansible_check_mode": false,
        "ansible_connection": "network_cli",
```

```
"ansible_diff_mode": false,
"ansible_facts": {
    "net_all_ipv4_addresses": [
        "10.0.0.5",
        "172.16.1.20",
        "192.168.0.1"
    ],
    "net_all_ipv6_addresses": [],
    "net_filesystems": [
        "flash0:"
    ],
    "net_gather_subset": [
        "hardware",
        "default",
        "interfaces"
    ],
    "net_hostname": "iosv-1",
    "net_image": "flash0:/vios-adventerprisek9-m",
    "net_interfaces": {
        "GigabitEthernet0/0": {
            "bandwidth": 1000000,
            "description": "OOB Management",
            "duplex": "Full",
            "ipv4": [
                {
                    "address": "172.16.1.20",
                    "subnet": "24"
                }
```
[skip]

The network facts module in Ansible 2.5 was a big step forward in streamlining your workflow and brought it on par with other server modules.

Network module conditional

Let's take a look at another network device conditional example by using the comparison keyword we saw at the beginning of this chapter. We can take advantage of the fact that both IOSv and Arista EOS provide the outputs in JSON format for the `show` commands. For example, we can check the status of the interface:

```
arista1#sh interfaces ethernet 1/3 | json
{
  "interfaces": {
  "Ethernet1/3": {
  "interfaceStatistics": {
<skip>
```

The Python Automation Framework – Beyond Basics

```
        "outPktsRate": 0.0
    },
    "name": "Ethernet1/3",
    "interfaceStatus": "disabled",
    "autoNegotiate": "off",
    <skip>
}
arista1#
```

If we have an operation that we want to perform and it depends on `Ethernet1/3` being disabled in order to have no user impact, such as to ensure no users are actively connected to `Ethernet1/3`, we can use the following tasks in the `chapter5_3.yml` playbook. It uses the `eos_command` module to gather the interface state output, and checks the interface status using the `waitfor` and `eq` keywords before proceeding to the next task:

```
   <skip>
    tasks:
      - name: "sh int ethernet 1/3 | json"
        eos_command:
          commands:
            - "show interface ethernet 1/3 | json"
          provider: "{{ cli }}"
          waitfor:
            - "result[0].interfaces.Ethernet1/3.interfaceStatus eq disabled"
        register: output
      - name: show output
        debug:
          msg: "Interface Disabled, Safe to Proceed"
```

Upon the condition being met, the second task will be executed:

```
TASK [sh int ethernet 1/3 | json]
*********************************************
ok: [arista1]

TASK [show output]
***************************************************************
ok: [arista1] => {
  "msg": "Interface Disabled, Safe to Proceed"
}
```

If the interface is active, an error will be given as follows following the first task:

```
TASK [sh int ethernet 1/3 | json]
*********************************************
fatal: [arista1]: FAILED! => {"changed": false, "commands": ["show interface ethernet 1/3 | json | json"], "failed": true, "msg":
```

```
"matched error in response: show interface ethernet 1/3 | json |
jsonrn% Invalid input (privileged mode required)rn*******1>"}
  to retry, use: --limit
@/home/echou/Master_Python_Networking/Chapter5/chapter5_3.retry

PLAY RECAP
*****************************************************************
arista1    : ok=0    changed=0    unreachable=0    failed=1
```

Check out the other conditions such as `contains`, `greater than`, and `less than`, as they fit into your situation.

Ansible loops

Ansible provides a number of loops in the playbook, such as standard loops, looping over files, subelements, do-until, and many more. In this section, we will look at two of the most commonly used loop forms: standard loops and looping over hash values.

Standard loops

Standard loops in playbooks are often used to easily perform similar tasks multiple times. The syntax for standard loops is very easy: the `{{ item }}` variable is the placeholder looping over the `with_items` list. For example, take a look at the following section in the `chapter5_4.yml` playbook:

```
tasks:
  - name: echo loop items
    command: echo {{ item }}
    with_items: ['r1', 'r2', 'r3', 'r4', 'r5']
```

It will loop over the five list items with the same `echo` command:

```
TASK [echo loop items]
***********************************************************
changed: [192.168.199.185] => (item=r1)
changed: [192.168.199.185] => (item=r2)
changed: [192.168.199.185] => (item=r3)
changed: [192.168.199.185] => (item=r4)
changed: [192.168.199.185] => (item=r5)
```

We will combine the standard loop with the network command module in the `chapter5_5.yml` playbook to add multiple VLANs to the device:

```
tasks:
  - name: add vlans
    eos_config:
      lines:
          - vlan {{ item }}
      provider: "{{ cli }}"
    with_items:
        - 100
        - 200
        - 300
```

The `with_items` list can also be read from a variable, which gives greater flexibility to the structure of your playbook:

```
vars:
  vlan_numbers: [100, 200, 300]
<skip>
tasks:
  - name: add vlans
    eos_config:
      lines:
          - vlan {{ item }}
      provider: "{{ cli }}"
    with_items: "{{ vlan_numbers }}"
```

The standard loop is a great time saver when it comes to performing redundant tasks in a playbook. It also makes the playbook more readable by reducing the lines required for the task.

In the next section, we will take a look at looping over dictionaries.

Looping over dictionaries

Looping over a simple list is nice. However, we often have an entity with more than one attribute associated with it. If you think about the `vlan` example in the last section, each `vlan` would have several unique attributes to it, such as the `vlan` description, the gateway IP address, and possibly others. Oftentimes, we can use a dictionary to represent the entity to incorporate multiple attributes to it.

Let's expand on the vlan example in the last section for a dictionary example in chapter5_6.yml. We defined the dictionary values for three vlans, each with a nested dictionary for the description and the IP address:

```
<skip>
vars:
  cli:
    host: "{{ ansible_host }}"
    username: "{{ username }}"
    password: "{{ password }}"
    transport: cli
  vlans: {
    "100": {"description": "floor_1", "ip": "192.168.10.1"},
    "200": {"description": "floor_2", "ip": "192.168.20.1"}
    "300": {"description": "floor_3", "ip": "192.168.30.1"}
  }
```

We can configure the first task, add vlans, by using the key of the each of items as the vlan number:

```
tasks:
  - name: add vlans
    nxos_config:
      lines:
        - vlan {{ item.key }}
      provider: "{{ cli }}"
    with_dict: "{{ vlans }}"
```

We can proceed with configuring the vlan interfaces. Note that we use the parents parameter to uniquely identify the section the commands should be checked against. This is due to the fact that the description and the IP address are both configured under the interface vlan <number> subsection in the configuration:

```
  - name: configure vlans
    nxos_config:
      lines:
        - description {{ item.value.name }}
        - ip address {{ item.value.ip }}/24
      provider: "{{ cli }}"
      parents: interface vlan {{ item.key }}
    with_dict: "{{ vlans }}"
```

Upon execution, you will see the dictionary being looped through:

```
TASK [configure vlans]
******************************************************
changed: [nxos-r1] => (item={'key': u'300', 'value': {u'ip':
```

```
u'192.168.30.1', u'name': u'floor_3'}})
changed: [nxos-r1] => (item={'key': u'200', 'value': {u'ip':
u'192.168.20.1', u'name': u'floor_2'}})
changed: [nxos-r1] => (item={'key': u'100', 'value': {u'ip':
u'192.168.10.1', u'name': u'floor_1'}})
```

Let's check if the intended configuration is applied to the device:

```
nx-osv-1# sh run | i vlan
<skip>
vlan 1,10,100,200,300
nx-osv-1#

nx-osv-1# sh run | section "interface Vlan100"
interface Vlan100
  description floor_1
  ip address 192.168.10.1/24
nx-osv-1#
```

For more loop types of Ansible, feel free to check out the documentation (http://docs.ansible.com/ansible/playbooks_loops.html).

Looping over dictionaries takes some practice the first few times you use them. But just like standard loops, looping over dictionaries will be an invaluable tool in your tool belt.

Templates

For as long as I can remember, working as a network engineer, I have always used a kind of network template. In my experience, many of the network devices have sections of the network configuration that are identical, especially if these devices serve the same role in the network.

Most of the time, when we need to provision a new device, we use the same configuration in the form of a template, replace the necessary fields, and copy the file over to the new device. With Ansible, you can automate all of the work by using the template module (http://docs.ansible.com/ansible/template_module.html).

The base template file we are using utilizes the Jinja2 template language (http://jinja.pocoo.org/docs/). We briefly discussed the Jinja2 templating language in Chapter 4, *The Python Automation Framework – Ansible Basics*, and we will look at it a bit more here. Just like Ansible, Jinja2 has its own syntax and method of doing loops and conditionals; fortunately, we just need to know the very basics of it for our purpose. The Ansible template is an important tool that we will be using in our daily task, and we will spend more of this section exploring it. We will learn the syntax by gradually building up our playbook from simple to more complex.

The basic syntax for template usage is very simple; you just need to specify the source file and the destination location that you want to copy it to.

We will create an empty file for now:

```
$ touch file1
```

Then, we will use the following playbook to copy `file1` to `file2`. Note that the playbook is executed on the control machine only. Next, we will specify the path of both the source and destination files as arguments for the `template` module:

```
---
- name: Template Basic
  hosts: localhost

  tasks:
    - name: copy one file to another
      template:
        src=./file1
        dest=./file2
```

We do not need to specify a host file during playbook execution since the localhost is available by default. However, you will get a warning:

```
$ ansible-playbook chapter5_7.yml
 [WARNING]: provided hosts list is empty, only localhost is available
<skip>
TASK [copy one file to another]
**************************************************

changed: [localhost]
<skip>
```

The Python Automation Framework – Beyond Basics

The source file can have any extension, but since they are processed through the Jinja2 template engine, let's create a text file called `nxos.j2` as the template source. The template will follow the Jinja2 convention of using double curly brace to specify the variables:

```
hostname {{ item.value.hostname }}
feature telnet
feature ospf
feature bgp
feature interface-vlan

username {{ item.value.username }} password {{ item.value.password }} role network-operator
```

The Jinja2 template

Let's also modify the playbook accordingly. In `chapter5_8.yml`, we will make the following changes:

1. Change the source file to `nxos.j2`
2. Change the destination file to be a variable
3. Provide the variable values as a dictionary that we will substitute in the template:

```
---
- name: Template Looping
  hosts: localhost

  vars:
    nexus_devices: {
      "nx-osv-1": {"hostname": "nx-osv-1", "username": "cisco", "password": "cisco"}
    }

  tasks:
    - name: create router configuration files
      template:
        src=./nxos.j2
        dest=./{{ item.key }}.conf
      with_dict: "{{ nexus_devices }}"
```

After running the playbook, you will find the destination file called `nx-osv-1.conf` with the values filled in and ready to be used:

```
$ cat nx-osv-1.conf
hostname nx-osv-1
```

```
feature telnet
feature ospf
feature bgp
feature interface-vlan

username cisco password cisco role network-operator
```

Jinja2 loops

We can also loop through a list as well as a dictionary in Jinja2. We will use both as loops in `nxos.j2`:

```
{% for vlan_num in item.value.vlans %}
vlan {{ vlan_num }}
{% endfor %}

{% for vlan_interface in item.value.vlan_interfaces %}
interface {{ vlan_interface.int_num }}
  ip address {{ vlan_interface.ip }}/24
{% endfor %}
```

Provide the additional list and dictionary variables in the `chapter5_8.yml` playbook:

```
vars:
  nexus_devices: {
    "nx-osv-1": {
    "hostname": "nx-osv-1",
    "username": "cisco",
    "password": "cisco",
    "vlans": [100, 200, 300],
    "vlan_interfaces": [
        {"int_num": "100", "ip": "192.168.10.1"},
        {"int_num": "200", "ip": "192.168.20.1"},
        {"int_num": "300", "ip": "192.168.30.1"}
      ]
    }
  }
```

Run the playbook, and you will see the configuration for both `vlan` and `vlan_interfaces` filled in on the router config.

The Jinja2 conditional

Jinja2 also supports an `if` conditional check. Let's add this field in for turning on the netflow feature for certain devices. We will add the following to the `nxos.j2` template:

```
{% if item.value.netflow_enable %}
feature netflow
{% endif %}
```

We will list out the difference in the playbook:

```
vars:
  nexus_devices: {
  <skip>
          "netflow_enable": True
  <skip>
  }
```

The last step we will undertake is to make `nxos.j2` more scalable by placing the `vlan` interface section inside of a `true-false` conditional check. In the real world, more often than not, we will have multiple devices with knowledge of the `vlan` information, but only one device as the gateway for client hosts:

```
{% if item.value.l3_vlan_interfaces %}
{% for vlan_interface in item.value.vlan_interfaces %}
interface {{ vlan_interface.int_num }}
 ip address {{ vlan_interface.ip }}/24
{% endfor %}
{% endif %}
```

We will also add a second device, called `nx-osv-2`, in the playbook:

```
vars:
  nexus_devices: {
  <skip>
    "nx-osv-2": {
      "hostname": "nx-osv-2",
      "username": "cisco",
      "password": "cisco",
      "vlans": [100, 200, 300],
      "l3_vlan_interfaces": False,
      "netflow_enable": False
    }
  <skip>
}
```

We are now ready to run our playbook:

```
$ ansible-playbook chapter5_8.yml
 [WARNING]: provided hosts list is empty, only localhost is available. Note
that the implicit localhost does not match 'all'


PLAY [Template Looping] ****************************************************

TASK [Gathering Facts] *****************************************************
ok: [localhost]

TASK [create router configuration files] ***********************************
ok: [localhost] => (item={'value': {u'username': u'cisco', u'password':
u'cisco', u'hostname': u'nx-osv-2', u'netflow_enable': False, u'vlans':
[100, 200, 300], u'l3_vlan_interfaces': False}, 'key': u'nx-osv-2'})
ok: [localhost] => (item={'value': {u'username': u'cisco', u'password':
u'cisco', u'hostname': u'nx-osv-1', u'vlan_interfaces': [{u'int_num':
u'100', u'ip': u'192.168.10.1'}, {u'int_num': u'200', u'ip':
u'192.168.20.1'}, {u'int_num': u'300', u'ip': u'192.168.30.1'}],
u'netflow_enable': True, u'vlans': [100, 200, 300], u'l3_vlan_interfaces':
True}, 'key': u'nx-osv-1'})

PLAY RECAP *****************************************************************
localhost     : ok=2    changed=0    unreachable=0    failed=0
```

Let's check the differences in the two configuration files to make sure that the conditional changes are taking place:

```
$ cat nx-osv-1.conf
hostname nx-osv-1

feature telnet
feature ospf
feature bgp
feature interface-vlan

feature netflow

username cisco password cisco role network-operator

vlan 100
vlan 200
vlan 300
```

```
interface 100
  ip address 192.168.10.1/24
interface 200
  ip address 192.168.20.1/24
interface 300
  ip address 192.168.30.1/24

$ cat nx-osv-2.conf
hostname nx-osv-2

feature telnet
feature ospf
feature bgp
feature interface-vlan


username cisco password cisco role network-operator

vlan 100
vlan 200
vlan 300
```

Neat, huh? This can certainly save us a ton of time for something that required repeated copy and paste before. Personally, the template module was a big game changer for me. This module alone was enough to motivate me to learn and use Ansible a few years ago.

Our playbook is getting kind of long. In the next section, we will see how we can optimize the playbook by offloading the variable files into groups and directories.

Group and host variables

Note that, in the previous playbook, `chapter5_8.yml`, we have repeated ourselves in the username and password variables for the two devices under the `nexus_devices` variable:

```
vars:
  nexus_devices: {
    "nx-osv-1": {
      "hostname": "nx-osv-1",
      "username": "cisco",
      "password": "cisco",
      "vlans": [100, 200, 300],
    <skip>
    "nx-osv-2": {
      "hostname": "nx-osv-2",
      "username": "cisco",
```

```
            "password": "cisco",
            "vlans": [100, 200, 300],
<skip>
```

This is not ideal. If we ever need to update the username and password values, we will need to remember to update at two locations. This increases the management burden as well as the chances of making mistakes. For a best practice, Ansible suggests that we use the `group_vars` and `host_vars` directories to separate out the variables.

 For more Ansible best practices, check out http://docs.ansible.com/ansible/playbooks_best_practices.html.

Group variables

By default, Ansible will look for group variables in the same directory as the playbook called `group_vars` for variables that can be applied to the group. By default, it will look for the filename that matches the group name in the inventory file. For example, if we have a group called [nexus-devices] in the inventory file, we can have a file under `group_vars` named `nexus-devices` to house all the variables that can be applied to the group.

We can also use a special file named `all` to include variables applied to all the groups.

We will utilize this feature for our username and password variables. First, we will create the `group_vars` directory:

```
$ mkdir group_vars
```

Then, we can create a YAML file called `all` to include the username and password:

```
$ cat group_vars/all
---
username: cisco
password: cisco
```

We can then use variables for the playbook:

```
    vars:
      nexus_devices: {
        "nx-osv-1": {
           "hostname": "nx-osv-1",
           "username": "{{ username }}",
           "password": "{{ password }}",
           "vlans": [100, 200, 300],
```

```
    <skip>
     "nx-osv-2": {
      "hostname": "nx-osv-2",
      "username": "{{ username }}",
      "password": "{{ password }}",
      "vlans": [100, 200, 300],
    <skip>
```

Host variables

We can further separate out the host variables in the same format as the group variables. This was how we were able to apply the variables in the Ansible 2.5 playbook examples in Chapter 4, *The Python Automation Framework – Ansible Basics*, and earlier in this chapter:

```
$ mkdir host_vars
```

In our case, we execute the commands on the localhost, and so the file under host_vars should be named accordingly, such as host_vars/localhost. In our host_vars/localhost file, we can also keep the variables declared in group_vars:

```
$ cat host_vars/localhost
---
"nexus_devices":
  "nx-osv-1":
    "hostname": "nx-osv-1"
    "username": "{{ username }}"
    "password": "{{ password }}"
    "vlans": [100, 200, 300]
    "l3_vlan_interfaces": True
    "vlan_interfaces": [
        {"int_num": "100", "ip": "192.168.10.1"},
        {"int_num": "200", "ip": "192.168.20.1"},
        {"int_num": "300", "ip": "192.168.30.1"}
    ]
    "netflow_enable": True

  "nx-osv-2":
    "hostname": "nx-osv-2"
    "username": "{{ username }}"
    "password": "{{ password }}"
    "vlans": [100, 200, 300]
    "l3_vlan_interfaces": False
    "netflow_enable": False
```

After we separate out the variables, the playbook now becomes very lightweight and only consists of the logic of our operation:

```
$ cat chapter5_9.yml
---
- name: Ansible Group and Host Variables
  hosts: localhost

  tasks:
    - name: create router configuration files
      template:
        src=./nxos.j2
        dest=./{{ item.key }}.conf
      with_dict: "{{ nexus_devices }}"
```

The `group_vars` and `host_vars` directories not only decrease our operations overhead, they can also help with securing the files by allowing us to encrypt the sensitive information with Ansible Vault, which we will look at next.

The Ansible Vault

As you can see from the previous section, in most cases, the Ansible variable provides sensitive information such as a username and password. It would be a good idea to put some security measures around the variables so that we can safeguard against them. The Ansible Vault (https://docs.ansible.com/ansible/2.5/user_guide/vault.html) provides encryption for files so they appear in plaintext.

All Ansible Vault functions start with the `ansible-vault` command. You can manually create an encrypted file via the create option. You will be asked to enter a password. If you try to view the file, you will find that the file is not in clear text. If you have downloaded the book example, the password I used was just the word *password*:

```
$ ansible-vault create secret.yml
Vault password: <password>

$ cat secret.yml
$ANSIBLE_VAULT;1.1;AES256
3365646264623739623266353263613236393236353536306466656564303532613837376 23
<skip>6535373338373838636365303564646230323334323861393033335663262
3962
```

The Python Automation Framework – Beyond Basics

To edit or view an encrypted file, we will use the `edit` option for edit or view the file via the `view` option:

```
$ ansible-vault edit secret.yml
Vault password:

$ ansible-vault view secret.yml
Vault password:
```

Let's encrypt the `group_vars/all and host_vars/localhost` variable files:

```
$ ansible-vault encrypt group_vars/all host_vars/localhost
Vault password:
Encryption successful
```

Now, when we run the playbook, we will get a decryption failed error message:

```
ERROR! Decryption failed on
/home/echou/Master_Python_Networking/Chapter5/Vaults/group_vars/all
```

We will need to use the `--ask-vault-pass` option when we run the playbook:

```
$ ansible-playbook chapter5_10.yml --ask-vault-pass
Vault password:
```

The decryption will happen in memory for any Vault-encrypted files that are accessed.

Prior to Ansible 2.4, Ansible Vault required all the files to be encrypted with the same password. Since Ansible 2.4 and later, you can use vault ID to supply a different password file (https://docs.ansible.com/ansible/2.5/user_guide/vault.html#multiple-vault-passwords).

We can also save the password in a file and make sure that the specific file has restricted permission:

```
$ chmod 400 ~/.vault_password.txt
$ ls -lia ~/.vault_password.txt
809496 -r--------  1 echou echou 9 Feb 18 12:17 /home/echou/.vault_password.txt
```

We can then execute the playbook with the `--vault-password-file` option:

```
$ ansible-playbook chapter5_10.yml --vault-password-file
~/.vault_password.txt
```

We can also encrypt just a string and embed the encrypted string inside of the playbook by using the `encrypt_string` option (https://docs.ansible.com/ansible/2.5/user_guide/vault.html#use-encrypt-string-to-create-encrypted-variables-to-embed-in-yaml):

```
$ ansible-vault encrypt_string
New Vault password:
Confirm New Vault password:
Reading plaintext input from stdin. (ctrl-d to end input)
new_user_password
!vault |
          $ANSIBLE_VAULT;1.1;AES256
61636438643839326262313962356161353965666438383464333832396662383634373737
36132613466323262386131333838353461386530386461636438306236365393665316133
61646264383165333266326364373436386666363646463656361626530366562636463656
23166356436462323135663163663331320a62356361326639333165393962663962306630303030
3761656435633966663343761303032663333643836626462646646366138323666376239656653
3623233353832

Encryption successful
```

The string can then be placed in the playbook file as a variable. In the next section, we will optimize our playbook even further with `include` and `roles`.

The Ansible include and roles

The best way to handle complex tasks is to break them down into smaller pieces. Of course, this approach is common in both Python and network engineering. In Python, we break complicated code into functions, classes, modules, and packages. In networking, we also break large networks into sections such as racks, rows, clusters, and datacenters. In Ansible, we can use `roles` and `includes` to segment and organize a large playbook into multiple files. Breaking up a large Ansible playbook simplifies the structure as each of the files focuses on fewer tasks. It also allows the sections of the playbook to be reused.

The Ansible include statement

As the playbook grows in size, it will eventually become obvious that many of the tasks and plays can be shared across different playbooks. The Ansible `include` statement is similar to many Linux configuration files that just tell the machine to extend the file the same way as if the file was directly written in. We can use an `include` statement for both playbooks and tasks. Here, we will look at a simple example of extending our task.

The Python Automation Framework – Beyond Basics

Let's assume that we want to show outputs for two different playbooks. We can make a separate YAML file called `show_output.yml` as an additional task:

```
---
- name: show output
    debug:
       var: output
```

Then, we can reuse this task in multiple playbooks, such as in `chapter5_11_1.yml`, which looks largely identical to the last playbook with the exception of registering the output and the include statement at the end:

```
---
- name: Ansible Group and Host Varibles
  hosts: localhost

  tasks:
    - name: create router configuration files
      template:
         src=./nxos.j2
         dest=./{{ item.key }}.conf
      with_dict: "{{ nexus_devices }}"
      register: output

    - include: show_output.yml
```

Another playbook, `chapter5_11_2.yml`, can reuse `show_output.yml` in the same way:

```
---
- name: show users
  hosts: localhost

  tasks:
    - name: show local users
      command: who
      register: output

    - include: show_output.yml
```

Note that both playbooks use the same variable name, `output`, because in `show_output.yml`, we hard coded the variable name for simplicity. You can also pass variables into the included file.

Ansible roles

Ansible roles separate the logical function with a physical host to fit your network better. For example, you can construct roles such as spines, leafs, core, as well as Cisco, Juniper, and Arista. The same physical host can belong to multiple roles; for example, a device can belong to both Juniper and the core. This flexibility allows us to perform operations, such as upgrade all Juniper devices, without worrying about the device's location in the layer of the network.

Ansible roles can automatically load certain variables, tasks, and handlers based on a known file infrastructure. The key is that this is a known file structure that we automatically include. In fact, you can think of roles as pre-made `include` statements by Ansible.

The Ansible playbook role documentation (http://docs.ansible.com/ansible/playbooks_roles.html#roles) describes a list of role directories that we can configure. We do not need to use all of them. In our example, we will only modify the `tasks and the vars` folders. However, it is good to know all of the available options in the Ansible role directory structure.

The following is what we will use as an example for our roles:

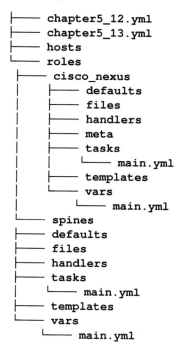

```
├── chapter5_12.yml
├── chapter5_13.yml
├── hosts
└── roles
    ├── cisco_nexus
    │   ├── defaults
    │   ├── files
    │   ├── handlers
    │   ├── meta
    │   ├── tasks
    │   │   └── main.yml
    │   ├── templates
    │   └── vars
    │       └── main.yml
    └── spines
        ├── defaults
        ├── files
        ├── handlers
        ├── tasks
        │   └── main.yml
        ├── templates
        └── vars
            └── main.yml
```

You can see that, at the top level, we have the hosts file as well as the playbooks. We also have a folder named `roles`. Inside the folder, we have two roles defined: `cisco_nexus` and `spines`. Most of the subfolders under the roles were empty, with the exception of the `tasks` and `vars` folders. There is a file named `main.yml` inside each of them. This is the default behavior: the `main.yml` file is your entry point that is automatically included in the playbook when you specify the role in the playbook. If you need to break out additional files, you can use the include statement in the `main.yml` file.

Here is our scenario:

- We have two Cisco Nexus devices, `nxos-r1` and `nxos-r2`. We will configure the logging server as well as the log link-status for all of them, utilizing the `cisco_nexus` role for them.
- In addition, nxos-r1 is also a spine device, where we will want to configure more verbose logging, perhaps because spines are at a more critical position within our network.

For our `cisco_nexus` role, we have the following variables in `roles/cisco_nexus/vars/main.yml`:

```
---
cli:
  host: "{{ ansible_host }}"
  username: cisco
  password: cisco
  transport: cli
```

We have the following configuration tasks in `roles/cisco_nexus/tasks/main.yml`:

```
---
- name: configure logging parameters
  nxos_config:
    lines:
      - logging server 191.168.1.100
      - logging event link-status default
    provider: "{{ cli }}"
```

Our playbook is extremely simple, as it just needs to specify the hosts that we would like to configure according to `cisco_nexus role`:

```
---
- name: playbook for cisco_nexus role
  hosts: "cisco_nexus"
  gather_facts: false
  connection: local
```

```
roles:
  - cisco_nexus
```

When you run the playbook, the playbook will include the tasks and variables defined in the `cisco_nexus` role and configure the devices accordingly.

For our `spine` role, we will have an additional task of more verbose logging in `roles/spines/tasks/mail.yml`:

```
---
- name: change logging level
  nxos_config:
    lines:
      - logging level local7 7
    provider: "{{ cli }}"
```

In our playbook, we can specify that it contains both the role of `cisco_nexus` as well as `spines`:

```
---
- name: playbook for spine role
  hosts: "spines"
  gather_facts: false
  connection: local

  roles:
    - cisco_nexus
    - spines
```

When we include both roles in this order, the `cisco_nexus` role tasks will be executed, followed by the spines role:

```
TASK [cisco_nexus : configure logging parameters]
*******************************
changed: [nxos-r1]

TASK [spines : change logging level]
*********************************************
ok: [nxos-r1]
```

Ansible roles are flexible and scalable – just like Python functions and classes. Once your code grows beyond a certain level, it is almost always a good idea to break it into smaller pieces for maintainability.

The Python Automation Framework – Beyond Basics

 You can find more examples of roles in the Ansible examples Git repository at `https://github.com/ansible/ansible-examples`.

Ansible Galaxy (`https://docs.ansible.com/ansible/latest/reference_appendices/galaxy.html`) is a free community site for finding, sharing, and collaborating on roles. You can see an example of the Juniper networks supplied by the Ansible role on Ansible Galaxy:

JUNOS Role on Ansible Galaxy (`https://galaxy.ansible.com/Juniper/junos`)

In the next section, we will take a look at how to write our own custom Ansible module.

Writing your own custom module

By now, you may get the feeling that network management in Ansible is largely dependent on finding the right module for your task. There is certainly a lot of truth in that logic. Modules provide a way to abstract the interaction between the managed host and the control machine; they allow us to focus on the logic of our operations. Up to this point, we have seen the major vendors providing a wide range of modules for Cisco, Juniper, and Arista.

[174]

Use the Cisco Nexus modules as an example, besides specific tasks such as managing the BGP neighbor (nxos_bgp) and the aaa server (nxos_aaa_server). Most vendors also provide ways to run arbitrary show (nxos_config) and configuration commands (nxos_config). This generally covers most of our use cases.

Starting with Ansible 2.5, there is also the streamline naming and usage of network facts modules.

What if the device you are using does not currently have the module for the task that you are looking for? In this section, we will look at several ways that we can remedy this situation by writing our own custom module.

The first custom module

Writing a custom module does not need to be complicated; in fact, it doesn't even need to be in Python. But since we are already familiar with Python, we will use Python for our custom modules. We are assuming that the module is what we will be using ourselves and our team without submitting back to Ansible, therefore we will ignore some of the documentation and formatting for the time being.

If you are interested in developing modules that can be submitted upstream to Ansible, please consult the developing modules guide from Ansible (https://docs.ansible.com/ansible/latest/dev_guide/developing_modules.html).

By default, if we create a folder named library in the same directory as the playbook, Ansible will include the directory in the module search path. Therefore, we can put our custom module in the directory and we will be able to use it in our playbook. The requirement for the custom module is very simple: all the module needs is to return a JSON output to the playbook.

Recall that in Chapter 3, *APIs and Intent-Driven Networking*, we used the following NXAPI Python script to communicate to the NX-OS device:

```
import requests
import json

url='http://172.16.1.142/ins'
switchuser='cisco'
switchpassword='cisco'
```

[175]

```
myheaders={'content-type':'application/json-rpc'}
payload=[
  {
    "jsonrpc": "2.0",
    "method": "cli",
    "params": {
      "cmd": "show version",
      "version": 1.2
    },
    "id": 1
  }
]
response = requests.post(url,data=json.dumps(payload),
headers=myheaders,auth=(switchuser,switchpassword)).json()

print(response['result']['body']['sys_ver_str'])
```

When we executed it, we simply received the system version. We can simply modify the last line to be a JSON output, as shown in the following code:

```
version = response['result']['body']['sys_ver_str']
print json.dumps({"version": version})
```

We will place this file under the `library` folder:

```
$ ls -a library/
. .. custom_module_1.py
```

In our playbook, we can then use the action plugin (https://docs.ansible.com/ansible/dev_guide/developing_plugins.html), chapter5_14.yml, to call this custom module:

```
---
- name: Your First Custom Module
  hosts: localhost
  gather_facts: false
  connection: local

  tasks:
    - name: Show Version
      action: custom_module_1
      register: output

    - debug:
        var: output
```

Note that, just like the `ssh` connection, we are executing the module locally with the module making API calls outbound. When you execute this playbook, you will get the following output:

```
$ ansible-playbook chapter5_14.yml
 [WARNING]: provided hosts list is empty, only localhost is available

PLAY [Your First Custom Module] ************************************

TASK [Show Version] ************************************************
ok: [localhost]

TASK [debug] *******************************************************
ok: [localhost] => {
 "output": {
 "changed": false,
 "version": "7.3(0)D1(1)"
 }
}

PLAY RECAP *********************************************************
localhost      : ok=2    changed=0    unreachable=0    failed=0
```

As you can see, you can write any module that is supported by API, and Ansible will happily take any returned JSON output.

The second custom module

Building upon the last module, let's utilize the common module boilerplate from Ansible that's stated in the module development documentation (http://docs.ansible.com/ansible/dev_guide/developing_modules_general.html). We will modify the last custom module and create `custom_module_2.py` to ingest inputs from the playbook.

First, we will import the boilerplate code from `ansible.module_utils.basic`:

```
from ansible.module_utils.basic import AnsibleModule
if __name__ == '__main__':
    main()
```

The Python Automation Framework – Beyond Basics

From there, we can define the main function where we will house our code. `AnsibleModule`, which we have already imported, provides lots of common code for handling returns and parsing arguments. In the following example, we will parse three arguments for `host`, `username`, and `password`, and make them required fields:

```
def main():
    module = AnsibleModule(
        argument_spec = dict(
        host = dict(required=True),
        username = dict(required=True),
        password = dict(required=True)
     )
  )
```

The values can then be retrieved and used in our code:

```
device = module.params.get('host')
username = module.params.get('username')
password = module.params.get('password')

url='http://' + host + '/ins'
switchuser=username
switchpassword=password
```

Finally, we will follow the exit code and return the value:

```
module.exit_json(changed=False, msg=str(data))
```

Our new playbook, `chapter5_15.yml`, will look identical to the last playbook, except now we can pass values for different devices in the playbook:

```
tasks:
  - name: Show Version
    action: custom_module_1 host="172.16.1.142" username="cisco" password="cisco"
    register: output
```

When executed, this playbook will produce the exact same output as the last playbook. However, because we are using arguments in the custom module, the custom module can now be passed around for other people to use without them knowing the details of our module. They can write in their own username, password, and host IP in the playbook.

Of course, this is a functional but incomplete module. For one, we did not perform any error checking, nor did we provide any documentation for usage. However, it is a good demonstration of how easy it is to build a custom module. The additional benefit is that we saw how we can use an existing script that we already made and turn it into a custom Ansible module.

Summary

In this chapter, we covered a lot of ground. Building from our previous knowledge of Ansible, we expanded into more advanced topics such as conditionals, loops, and templates. We looked at how to make our playbook more scalable with host variables, group variables, include statements, and roles. We also looked at how to secure our playbook with the Ansible Vault. Finally, we used Python to make our own custom modules.

Ansible is a very flexible Python framework that can be used for network automation. It provides another abstraction layer separated from the likes of the Pexpect and API-based scripts. It is declarative in nature in that it is more expressive in terms of matching our intent. Depending on your needs and network environment, it might be the ideal framework that you can use to save time and energy.

In `Chapter 6`, *Network Security with Python*, we will look at network security with Python.

Network Security with Python

In my opinion, network security is a tricky topic to write about. The reason is not a technical one, but rather to do with setting up the right scope. The boundaries of network security are so wide that they touch all seven layers of the OSI model. From layer 1 of wiretapping to layer 4 of the transport protocol vulnerability, to layer 7 of man-in-the-middle spoofing, network security is everywhere. The issue is exacerbated by all the newly discovered vulnerabilities, which sometimes seem to be at a daily rate. This does not even include the human social engineering aspect of network security.

As such, in this chapter, I would like to set the scope for what we will discuss. As we have been doing up to this point, we will primarily focus on using Python for network device security at OSI layers 3 and 4. We will look at Python tools that we can use to manage individual network devices for security purposes, as well as using Python as a glue to connect different components. Hopefully, we can treat network security with a holistic view by using Python in different OSI layers.

In this chapter, we will take a look at the following topics:

- The lab setup
- Python Scapy for security testing
- Access lists
- Forensic analysis with Syslog and UFW using Python
- Other tools, such as a MAC address filter list, private VLAN, and Python IP table binding

The lab setup

The devices being used in this chapter are a bit different from the previous chapters. In the previous chapters, we were isolating a particular device by focusing on the topic at hand. For this chapter, we will use a few more devices in our lab in order to illustrate the function of the tools that we will be using. The connectivity and operating system information are important as they have ramifications regarding the security tools that we will show later in this chapter. For example, if we want to apply an access list to protect the server, we need to know what the topology looks like and which direction the client is making their connections from. The Ubuntu host connections are a bit different than what we have seen so far, so please make reference back to this lab section when you see the example later if needed.

We will be using the same Cisco VIRL tool with four nodes: two hosts and two network devices. If you need a refresher on Cisco VIRL, feel free to go back to Chapter 2, *Low-Level Network Device Interactions*, where we first introduced the tool:

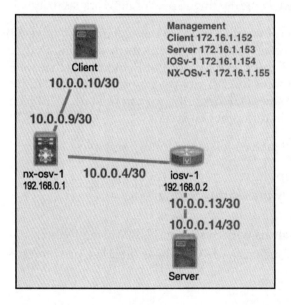

Lab topology

The IP addresses listed will be different in your own lab. They are listed here for an easy reference for the rest of the chapter.

Chapter 6

As illustrated, we will rename the host on the top as the client and the bottom host as the server. This is analogous to an internet client trying to access a corporate server within our network. We will again use the **Shared flat network** option for the management network to access the devices for the out-of-band management:

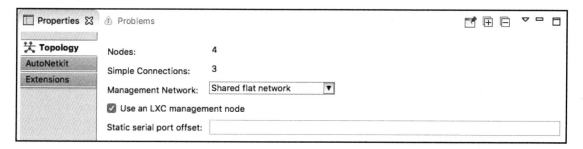

For the two switches, I will choose **Open Shortest Path First** (**OSPF**) as `IGP` and put both the devices in area `0`. By default, `BGP` is turned on and both the devices are using AS 1. From the configuration auto generation, the interfaces connected to the Ubuntu hosts are put into OSPF area `1`, so they will show up as inter-area routes. The NX-OSv configurations are shown here and the IOSv configuration and output are similar:

```
interface Ethernet2/1
  description to iosv-1
  no switchport
  mac-address fa16.3e00.0001
  ip address 10.0.0.6/30
  ip router ospf 1 area 0.0.0.0
  no shutdown

interface Ethernet2/2
  description to Client
  no switchport
  mac-address fa16.3e00.0002
  ip address 10.0.0.9/30
  ip router ospf 1 area 0.0.0.0
  no shutdown

nx-osv-1# sh ip route
<skip>
10.0.0.12/30, ubest/mbest: 1/0
  *via 10.0.0.5, Eth2/1, [110/41], 04:53:02, ospf-1, intra
192.168.0.2/32, ubest/mbest: 1/0
  *via 10.0.0.5, Eth2/1, [110/41], 04:53:02, ospf-1, intra
<skip>
```

[183]

The OSPF neighbor and the BGP output for NX-OSv are shown here, and the IOSv output is similar:

```
nx-osv-1# sh ip ospf neighbors
 OSPF Process ID 1 VRF default
 Total number of neighbors: 1
 Neighbor ID Pri State Up Time Address Interface
 192.168.0.2 1 FULL/DR 04:53:00 10.0.0.5 Eth2/1

nx-osv-1# sh ip bgp summary
BGP summary information for VRF default, address family IPv4 Unicast
BGP router identifier 192.168.0.1, local AS number 1
BGP table version is 5, IPv4 Unicast config peers 1, capable peers 1
2 network entries and 2 paths using 288 bytes of memory
BGP attribute entries [2/288], BGP AS path entries [0/0]
BGP community entries [0/0], BGP clusterlist entries [0/0]

Neighbor V AS MsgRcvd MsgSent TblVer InQ OutQ Up/Down State/PfxRcd
192.168.0.2 4 1 321 297 5 0 0 04:52:56 1
```

The hosts in our network are running Ubuntu 14.04, similar to the Ubuntu VM 16.04 that we have been using up to this point:

```
cisco@Server:~$ lsb_release -a
No LSB modules are available.
Distributor ID: Ubuntu
Description: Ubuntu 14.04.2 LTS
Release: 14.04
Codename: trusty
```

On both of the Ubuntu hosts, there are two network interfaces, `eth0` and `eth1`. `eth0` connects to the management network (`172.16.1.0/24`) while `eth1` connects to the network devices (`10.0.0.x/30`). The routes to the device loopback are directly connected to the network block, and the remote host network is statically routed to `eth1` with the default route going toward the management network:

```
cisco@Client:~$ route -n
Kernel IP routing table
Destination Gateway Genmask Flags Metric Ref Use Iface
0.0.0.0 172.16.1.2 0.0.0.0 UG 0 0 0 eth0
10.0.0.4 10.0.0.9 255.255.255.252 UG 0 0 0 eth1
10.0.0.8 0.0.0.0 255.255.255.252 U 0 0 0 eth1
10.0.0.8 10.0.0.9 255.255.255.248 UG 0 0 0 eth1
172.16.1.0 0.0.0.0 255.255.255.0 U 0 0 0 eth0
192.168.0.1 10.0.0.9 255.255.255.255 UGH 0 0 0 eth1
192.168.0.2 10.0.0.9 255.255.255.255 UGH 0 0 0 eth1
```

To verify the client-to-server path, let's ping and trace the route to make sure that traffic between our hosts is going through the network devices instead of the default route:

```
## Our server IP is 10.0.0.14
cisco@Server:~$ ifconfig
<skip>
eth1 Link encap:Ethernet HWaddr fa:16:3e:d6:83:02
     inet addr:10.0.0.14 Bcast:10.0.0.15 Mask:255.255.255.252

## From the client ping toward server
cisco@Client:~$ ping -c 1 10.0.0.14
PING 10.0.0.14 (10.0.0.14) 56(84) bytes of data.
64 bytes from 10.0.0.14: icmp_seq=1 ttl=62 time=6.22 ms

--- 10.0.0.14 ping statistics ---
1 packets transmitted, 1 received, 0% packet loss, time 0ms
rtt min/avg/max/mdev = 6.223/6.223/6.223/0.000 ms

## Traceroute from client to server
cisco@Client:~$ traceroute 10.0.0.14
traceroute to 10.0.0.14 (10.0.0.14), 30 hops max, 60 byte packets
 1  10.0.0.9 (10.0.0.9)  11.335 ms  11.745 ms  12.113 ms
 2  10.0.0.5 (10.0.0.5)  24.221 ms  41.635 ms  41.638 ms
 3  10.0.0.14 (10.0.0.14)  37.916 ms  38.275 ms  38.588 ms
cisco@Client:~$
```

Great! We have our lab; we are now ready to look at some security tools and measures using Python.

Python Scapy

Scapy (https://scapy.net) is a powerful Python-based interactive packet crafting program. Outside of some expensive commercial programs, very few tools can do what Scapy can do, to my knowledge. It is one of my favorite tools in Python.

The main advantage of Scapy is that it allows you to craft your own packet from the very basic level. In the words of Scapy's creator:

> "Scapy is a powerful interactive packet manipulation program. It is able to forge or decode packets of a wide number of protocols, send them on the wire, capture them, match requests and replies, and much more.... with most other tools, you won't build something the author did not imagine. These tools have been built for a specific goal and can't deviate much from it."

Let's take a look at the tool.

Installing Scapy

At the time of writing, Scapy 2.3.1 supported Python 2.7. Unfortunately, there were a few misfires regarding Python 3 support for Scapy and it is still relatively new for Scapy 2.3.3. For your environment, please feel free to try out Python 3 with version 2.3.3 and later. In this chapter, we will use Scapy 2.3.1 with Python 2.7. See the information sidebar if you would like to learn more about the reason behind the choice.

The long story for Python 3 support in Scapy is that there was an independent fork of Scapy from version 2.2.0 in 2015, aimed at supporting only Python 3. The project was named `Scapy3k`. The fork diverged from the main Scapy code base. If you read the first edition of this book, that was the information provided at the time of writing. There were confusions surrounding `python3-scapy` on PyPI and the official support of the Scapy code base. Our main purpose was to learn about Scapy in this chapter, and so therefore I made the choice to use an older, Python 2-based Scapy version.

In our lab, since we are crafting packet sources from the client to the destination server, Scapy needs to be installed on the client:

```
cisco@Client:~$ sudo apt-get update
cisco@Client:~$ sudo apt-get install git
cisco@Client:~$ git clone https://github.com/secdev/scapy
cisco@Client:~$ cd scapy/
cisco@Client:~/scapy$ sudo python setup.py install
```

Here is a quick test to make sure that the packages have been installed correctly:

```
cisco@Client:~/scapy$ python
Python 2.7.6 (default, Mar 22 2014, 22:59:56)
[GCC 4.8.2] on linux2
Type "help", "copyright", "credits" or "license" for more information.
>>> from scapy.all import *
```

Interactive examples

In our first example, we will craft an **Internet Control Message Protocol (ICMP)** packet on the client and send it to the server. On the server side, we will use `tcpdump` with a host filter to see the packet coming in:

```
## Client Side
cisco@Client:~/scapy$ sudo scapy
<skip>
Welcome to Scapy (2.3.3.dev274)
>>> send(IP(dst="10.0.0.14")/ICMP())
.
Sent 1 packets.
>>>

## Server Side
cisco@Server:~$ sudo tcpdump -i eth1 host 10.0.0.10
tcpdump: verbose output suppressed, use -v or -vv for full protocol decode
listening on eth1, link-type EN10MB (Ethernet), capture size 65535 bytes
02:45:16.400162 IP 10.0.0.10 > 10.0.0.14: ICMP echo request, id 0, seq 0, length 8
02:45:16.400192 IP 10.0.0.14 > 10.0.0.10: ICMP echo reply, id 0, seq 0, length 8
```

As you can see, it is very simple to craft a packet from Scapy. Scapy allows you to build the packet layer by layer using the slash (/) as the separator. The `send` function operates at the layer 3 level, which takes care of routing and layer 2 for you. There is also a `sendp()` alternative that operates at layer 2, which means you will need to specify the interface and link layer protocol.

Let's look at capturing the returned packet by using the send-request (`sr`) function. We are using a special variation of `sr`, called `sr1`, which only returns one packet that answers from the packet sent:

```
>>> p = sr1(IP(dst="10.0.0.14")/ICMP())
>>> p
<IP version=4L ihl=5L tos=0x0 len=28 id=26713 flags= frag=0L ttl=62 proto=icmp chksum=0x71 src=10.0.0.14 dst=10.0.0.10 options=[] |<ICMP type=echo-reply code=0 chksum=0xffff id=0x0 seq=0x0 |>>
```

One thing to note is that the `sr()` function itself returns a tuple containing answered and unanswered lists:

```
>>> p = sr(IP(dst="10.0.0.14")/ICMP())
>>> type(p)
<type 'tuple'>
```

```
## unpacking
>>> ans,unans = sr(IP(dst="10.0.0.14")/ICMP())
>>> type(ans)
<class 'scapy.plist.SndRcvList'>
>>> type(unans)
<class 'scapy.plist.PacketList'>
```

If we were to only take a look at the answered packet list, we can see it is another tuple containing the packet that we have sent as well as the returned packet:

```
>>> for i in ans:
...     print(type(i))
...
<type 'tuple'>
>>> for i in ans:
...     print i
...
(<IP frag=0 proto=icmp dst=10.0.0.14 |<ICMP |>>, <IP version=4L ihl=5L tos=0x0 len=28 id=27062 flags= frag=0L ttl=62 proto=icmp chksum=0xff13 src=10.0.0.14 dst=10.0.0.10 options=[] |<ICMP type=echo-reply code=0 chksum=0xffff id=0x0 seq=0x0 |>>)
```

Scapy also provides a layer 7 construct as well, such as a `DNS` query. In the following example, we are querying an open DNS server for the resolution of `www.google.com`:

```
>>> p = sr1(IP(dst="8.8.8.8")/UDP()/DNS(rd=1,qd=DNSQR(qname="www.google.com")))
>>> p
<IP version=4L ihl=5L tos=0x0 len=76 id=21743 flags= frag=0L ttl=128 proto=udp chksum=0x27fa src=8.8.8.8 dst=172.16.1.152 options=[] |<UDP sport=domain dport=domain len=56 chksum=0xc077 |<DNS id=0 qr=1L opcode=QUERY aa=0L tc=0L rd=1L ra=1L z=0L ad=0L cd=0L rcode=ok qdcount=1 ancount=1 nscount=0 arcount=0 qd=<DNSQR qname='www.google.com.' qtype=A qclass=IN |> an=<DNSRR rrname='www.google.com.' type=A rclass=IN ttl=299 rdata='172.217.3.164' |> ns=None ar=None |>>>
>>>
```

Sniffing

Scapy can also be used to easily capture packets on the wire:

```
>>> a = sniff(filter="icmp and host 172.217.3.164", count=5)
>>> a.show()
0000 Ether / IP / TCP 192.168.225.146:ssh > 192.168.225.1:50862 PA / Raw
0001 Ether / IP / ICMP 192.168.225.146 > 172.217.3.164 echo-request 0 / Raw
0002 Ether / IP / ICMP 172.217.3.164 > 192.168.225.146 echo-reply 0 / Raw
```

```
0003 Ether / IP / ICMP 192.168.225.146 > 172.217.3.164 echo-request 0 / Raw
0004 Ether / IP / ICMP 172.217.3.164 > 192.168.225.146 echo-reply 0 / Raw
>>>
```

We can look at the packets in some more detail, including the raw format:

```
>>> for i in a:
...     print i.show()
...
<skip>
###[ Ethernet ]###
 dst= <>
 src= <>
 type= 0x800
###[ IP ]###
 version= 4L
 ihl= 5L
 tos= 0x0
 len= 84
 id= 15714
 flags= DF
 frag= 0L
 ttl= 64
 proto= icmp
 chksum= 0xaa8e
 src= 192.168.225.146
 dst= 172.217.3.164
 options
###[ ICMP ]###
 type= echo-request
 code= 0
 chksum= 0xe1cf
 id= 0xaa67
 seq= 0x1
###[ Raw ]###
 load=
'xd6xbfxb1Xx00x00x00x00x1axdcnx00x00x00x00x00x10x11x12x13x14x15x16x17x18x19
x1ax1bx1cx1dx1ex1f !"#$%&'()*+,-./01234567'
None
```

We have seen the basic workings of Scapy. Let's move on and see how we can use Scapy for some of the common security testings.

The TCP port scan

The first step for any potential hackers is almost always trying to learn which service is open on the network, so they can concentrate their effort on the attack. Of course, we need to open certain ports in order to service our customer; that is part of the risk we need to accept. But we should also close any other open port that needlessly expose a larger attack surface. We can use Scapy to do a simple TCP open port scan to scan our own host.

We can send a `SYN` packet and see whether the server will return with `SYN-ACK`:

```
>>> p = sr1(IP(dst="10.0.0.14")/TCP(sport=666,dport=23,flags="S"))
>>> p.show()
###[ IP ]###
  version= 4L
  ihl= 5L
  tos= 0x0
  len= 40
  id= 25373
  flags= DF
  frag= 0L
  ttl= 62
  proto= tcp
  chksum= 0xc59b
  src= 10.0.0.14
  dst= 10.0.0.10
  options
###[ TCP ]###
  sport= telnet
  dport= 666
  seq= 0
  ack= 1
  dataofs= 5L
  reserved= 0L
  flags= RA
  window= 0
  chksum= 0x9907
  urgptr= 0
  options= {}
```

Note that, in the output here, the server is responding with a `RESET+ACK` for TCP port 23. However, TCP port 22 (SSH) is open; therefore, a `SYN-ACK` is returned:

```
>>> p = sr1(IP(dst="10.0.0.14")/TCP(sport=666,dport=22,flags="S"))
>>> p.show()
###[ IP ]###
  version= 4L
<skip>
```

```
  proto= tcp
  chksum= 0x28b5
  src= 10.0.0.14
  dst= 10.0.0.10
  options
###[ TCP ]###
  sport= ssh
  dport= 666
<skip>
  flags= SA
<skip>
```

We can also scan a range of destination ports from 20 to 22; note that we are using sr() for send-receive instead of the sr1() send-receive-one-packet variant:

```
>>> ans,unans =
sr(IP(dst="10.0.0.14")/TCP(sport=666,dport=(20,22),flags="S"))
>>> for i in ans:
...       print i
...
(<IP frag=0 proto=tcp dst=10.0.0.14 |<TCP sport=666 dport=ftp_data flags=S
|>>, <IP version=4L ihl=5L tos=0x0 len=40 id=4126 flags=DF frag=0L ttl=62
proto=tcp chksum=0x189b src=10.0.0.14 dst=10.0.0.10 options=[] |<TCP
sport=ftp_data dport=666 seq=0 ack=1 dataofs=5L reserved=0L flags=RA
window=0 chksum=0x990a urgptr=0 |>>)
(<IP frag=0 proto=tcp dst=10.0.0.14 |<TCP sport=666 dport=ftp flags=S |>>,
<IP version=4L ihl=5L tos=0x0 len=40 id=4127 flags=DF frag=0L ttl=62
proto=tcp chksum=0x189a src=10.0.0.14 dst=10.0.0.10 options=[] |<TCP
sport=ftp dport=666 seq=0 ack=1 dataofs=5L reserved=0L flags=RA window=0
chksum=0x9909 urgptr=0 |>>)
(<IP frag=0 proto=tcp dst=10.0.0.14 |<TCP sport=666 dport=ssh flags=S |>>,
<IP version=4L ihl=5L tos=0x0 len=44 id=0 flags=DF frag=0L ttl=62 proto=tcp
chksum=0x28b5 src=10.0.0.14 dst=10.0.0.10 options=[] |<TCP sport=ssh
dport=666 seq=4187384571 ack=1 dataofs=6L reserved=0L flags=SA window=29200
chksum=0xaaab urgptr=0 options=[('MSS', 1460)] |>>)
>>>
```

We can also specify a destination network instead of a single host. As you can see from the 10.0.0.8/29 block, hosts 10.0.0.9, 10.0.0.13, and 10.0.0.14 returned with SA, which corresponds to the two network devices and the host:

```
>>> ans,unans =
sr(IP(dst="10.0.0.8/29")/TCP(sport=666,dport=(22),flags="S"))
>>> for i in ans:
...       print(i)
...
(<IP frag=0 proto=tcp dst=10.0.0.9 |<TCP sport=666 dport=ssh flags=S |>>,
<IP version=4L ihl=5L tos=0x0 len=44 id=7304 flags= frag=0L ttl=64
```

```
 proto=tcp chksum=0x4a32 src=10.0.0.9 dst=10.0.0.10 options=[] |<TCP
 sport=ssh dport=666 seq=541401209 ack=1 dataofs=6L reserved=0L flags=SA
 window=17292 chksum=0xfd18 urgptr=0 options=[('MSS', 1444)] |>>,
 (<IP frag=0 proto=tcp dst=10.0.0.14 |<TCP sport=666 dport=ssh flags=S |>>,
 <IP version=4L ihl=5L tos=0x0 len=44 id=0 flags=DF frag=0L ttl=62 proto=tcp
 chksum=0x28b5 src=10.0.0.14 dst=10.0.0.10 options=[] |<TCP sport=ssh
 dport=666 seq=4222593330 ack=1 dataofs=6L reserved=0L flags=SA window=29200
 chksum=0x6a5b urgptr=0 options=[('MSS', 1460)] |>>,
 (<IP frag=0 proto=tcp dst=10.0.0.13 |<TCP sport=666 dport=ssh flags=S |>>,
 <IP version=4L ihl=5L tos=0x0 len=44 id=41992 flags= frag=0L ttl=254
 proto=tcp chksum=0x4ad src=10.0.0.13 dst=10.0.0.10 options=[] |<TCP
 sport=ssh dport=666 seq=2167267659 ack=1 dataofs=6L reserved=0L flags=SA
 window=4128 chksum=0x1252 urgptr=0 options=[('MSS', 536)] |>>)
```

Based on what we have learned so far, we can make a simple script for reusability, `scapy_tcp_scan_1.py`. We start with the suggested importing of `scapy` and the `sys` module for taking in arguments:

```
#!/usr/bin/env python2

from scapy.all import *
import sys
```

The `tcp_scan()` function is similar to what we have seen up to this point:

```
def tcp_scan(destination, dport):
    ans, unans =
sr(IP(dst=destination)/TCP(sport=666,dport=dport,flags="S"))
    for sending, returned in ans:
        if 'SA' in str(returned[TCP].flags):
            return destination + " port " + str(sending[TCP].dport) + " is open"
        else:
            return destination + " port " + str(sending[TCP].dport) + " is not open"
```

We can then acquire the input from arguments, and then call the `tcp_scan()` function in `main()`:

```
def main():
    destination = sys.argv[1]
    port = int(sys.argv[2])
    scan_result = tcp_scan(destination, port)
    print(scan_result)

if __name__ == "__main__":
    main()
```

Remember that access to the low-level network requires root access; therefore, our script needs to be executed as `sudo`:

```
cisco@Client:~$ sudo python scapy_tcp_scan_1.py "10.0.0.14" 23
<skip>
10.0.0.14 port 23 is not open
cisco@Client:~$ sudo python scapy_tcp_scan_1.py "10.0.0.14" 22
<skip>
10.0.0.14 port 22 is open
```

This was a relatively lengthy example of the TCP scan script, which demonstrated the power of crafting your own packet with Scapy. We tested out the steps in the interactive shell and finalized the usage with a simple script. Let's look at some more examples of Scapy's usage for security testing.

The ping collection

Let's say our network contains a mix of Windows, Unix, and Linux machines with users adding their own **Bring Your Own Device (BYOD)**; they may or may not support an ICMP ping. We can now construct a file with three types of common pings for our network, the ICMP, TCP, and UDP pings, in `scapy_ping_collection.py`:

```
#!/usr/bin/env python2

from scapy.all import *

def icmp_ping(destination):
    # regular ICMP ping
    ans, unans = sr(IP(dst=destination)/ICMP())
    return ans

def tcp_ping(destination, dport):
    # TCP SYN Scan
    ans, unans = sr(IP(dst=destination)/TCP(dport=dport,flags="S"))
    return ans

def udp_ping(destination):
    # ICMP Port unreachable error from closed port
    ans, unans = sr(IP(dst=destination)/UDP(dport=0))
    return ans
```

In this example, we will also use `summary()` and `sprintf()` for the output:

```
def answer_summary(answer_list):
  # example of lambda with pretty print
    answer_list.summary(lambda(s, r): r.sprintf("%IP.src% is alive"))
```

 If you were wondering why there is a lambda in the preceding `answer_summary()` function, it is a way to create a small anonymous function. Basically, it is a function without a name. More information on it can be found at https://docs.python.org/3.5/tutorial/controlflow.html#lambda-expressions.

We can then execute all three types of pings on the network in one script:

```
def main():
    print("** ICMP Ping **")
    ans = icmp_ping("10.0.0.13-14")
    answer_summary(ans)
    print("** TCP Ping **")
    ans = tcp_ping("10.0.0.13", 22)
    answer_summary(ans)
    print("** UDP Ping **")
    ans = udp_ping("10.0.0.13-14")
    answer_summary(ans)

if __name__ == "__main__":
    main()
```

At this point, hopefully, you will agree with me that, by having the ability to construct your own packet, you can be in charge of the type of operations and tests that you would like to run.

Common attacks

In this example, let's look at how we can construct our packet to conduct some of the classic attacks, such as *Ping of Death* (https://en.wikipedia.org/wiki/Ping_of_death) and *Land Attack* (https://en.wikipedia.org/wiki/Denial-of-service_attack). This is perhaps the network penetration tests that you previously had to pay for with a similar commercial software. With Scapy, you can conduct the test while maintaining full control as well as adding more tests in the future.

The first attack basically sends the destination host with a bogus IP header, such as the length of 2 and the IP version 3:

```
def malformed_packet_attack(host):
    send(IP(dst=host, ihl=2, version=3)/ICMP())
```

The `ping_of_death_attack` consists of the regular ICMP packet with a payload bigger than 65,535 bytes:

```
def ping_of_death_attack(host):
    # https://en.wikipedia.org/wiki/Ping_of_death
    send(fragment(IP(dst=host)/ICMP()/("X"*60000)))
```

The `land_attack` wants to redirect the client response back to the client itself and exhausts the host's resources:

```
def land_attack(host):
    # https://en.wikipedia.org/wiki/Denial-of-service_attack
    send(IP(src=host, dst=host)/TCP(sport=135,dport=135))
```

These are pretty old vulnerabilities or classic attacks that the modern operating system is no longer susceptible to. For our Ubuntu 14.04 host, none of the preceding attacks will bring it down. However, as more security issues are being discovered, Scapy is a great tool to start tests against our own network and host without having to wait for the impacted vendor to give you a validation tool. This is especially true for the zero-day (published without prior notification) attacks that seem to be more and more common on the internet.

Scapy resources

We have spent quite a bit of effort working with Scapy in this chapter. This is partially due to how highly I personally think of the tool. I hope you agree with me that Scapy is a great tool to keep in your toolset as a network engineer. The best part about Scapy is that it is constantly being developed with an engaged community of users.

 I would highly recommend at least going through the Scapy tutorial at http://scapy.readthedocs.io/en/latest/usage.html#interactive-tutorial, as well as any of the documentation that is of interest to you.

Access lists

The network access lists are usually the first line of defense against outside intrusions and attacks. Generally speaking, routers and switches process packets at a much faster rate than servers, because they utilize hardware such as **Ternary Content-Addressable Memory (TCAM)**. They do not need to see the application layer information, rather they just examine the layer 3 and layer 4 information, and decide whether the packets can be forwarded on or not. Therefore, we generally utilize network device access lists as the first step in safeguarding our network resources.

As a rule of thumb, we want to place access lists as close to the source (client) as possible. Inherently, we also trust the inside host and distrust the clients outside of our network boundary. The access list is therefore usually placed on the inbound direction on the external facing network interface(s). In our lab scenario, this means we will place an inbound access list at Ethernet2/2 that is directly connected to the client host.

If you are unsure of the direction and placement of the access list, a few points might help here:

- Think of the access list from the perspective of the network device
- Simplify the packets in terms of just source and destination IP and use one host as an example:
    - In our lab, traffic from our server will have a source IP of 10.0.0.14 with the destination IP of 10.0.0.10
    - The traffic from the client will have a source IP of 10.10.10.10 and the destination IP of 10.0.0.14

Obviously, every network is different and how the access list should be constructed depends on the services provided by your server. But as an inbound border access list, you should do the following:

- Deny RFC 3030 special-use address sources, such as 127.0.0.0/8
- Deny RFC 1918 space, such as 10.0.0.0/8
- Deny our own space as the source IP; in this case, 10.0.0.12/30
- Permit inbound TCP port 22 (SSH) and 80 (HTTP) to host 10.0.0.14
- Deny everything else

Implementing access lists with Ansible

The easiest way to implement this access list would be to use Ansible. We have already looked at Ansible in the last two chapters, but it is worth repeating the advantages of using Ansible in this scenario:

- **Easier management**: For a long access list, we are able to utilize the `include` statement to break it into more manageable pieces. The smaller pieces can then be managed by other teams or service owners.
- **Idempotency**: We can schedule the playbook at a regular interval and only the necessary changes will be made.
- **Each task is explicit**: We can separate the construct of the entries as well as apply the access list to the proper interface.
- **Reusability**: In the future, if we add additional external-facing interfaces, we just need to add the device to the list of devices for the access list.
- **Extensible**: You will notice that we can use the same playbook for constructing the access list and apply it to the right interface. We can start small and expand to separate playbooks in the future as needed.

The host file is pretty standard. For simplicity, we are putting the host variables directly in the inventory file:

```
[nxosv-devices]
nx-osv-1 ansible_host=172.16.1.155 ansible_username=cisco
ansible_password=cisco
```

We will declare the variables in the playbook for the time being:

```
---
- name: Configure Access List
  hosts: "nxosv-devices"
  gather_facts: false
  connection: local

  vars:
    cli:
      host: "{{ ansible_host }}"
      username: "{{ ansible_username }}"
      password: "{{ ansible_password }}"
      transport: cli
```

To save space, we will illustrate denying the RFC 1918 space only. Implementing the denial of RFC 3030 and our own space will be identical to the steps used for the RFC 1918 space. Note that we did not deny `10.0.0.0/8` in our playbook, because our configuration currently uses the `10.0.0.0` network for addressing. Of course, we could perform the single host permit first and deny `10.0.0.0/8` in a later entry, but in this example, we just choose to omit it:

```
tasks:
  - nxos_acl:
      name: border_inbound
      seq: 20
      action: deny
      proto: tcp
      src: 172.16.0.0/12
      dest: any
      log: enable
      state: present
      provider: "{{ cli }}"
  - nxos_acl:
      name: border_inbound
      seq: 40
      action: permit
      proto: tcp
      src: any
      dest: 10.0.0.14/32
      dest_port_op: eq
      dest_port1: 22
      state: present
      log: enable
      provider: "{{ cli }}"
  - nxos_acl:
      name: border_inbound
      seq: 50
      action: permit
      proto: tcp
      src: any
      dest: 10.0.0.14/32
      dest_port_op: eq
      dest_port1: 80
      state: present
      log: enable
      provider: "{{ cli }}"
  - nxos_acl:
      name: border_inbound
      seq: 60
      action: permit
      proto: tcp
```

```
        src: any
        dest: any
        state: present
        log: enable
        established: enable
        provider: "{{ cli }}"
    - nxos_acl:
        name: border_inbound
        seq: 1000
        action: deny
        proto: ip
        src: any
        dest: any
        state: present
        log: enable
        provider: "{{ cli }}"
```

Note that we are allowing the established connection sourcing from the server inside to be allowed back in. We use the final explicit `deny ip any any` statement as a high sequence number (`1000`), so we can insert any new entries later on.

We can then apply the access list to the right interface:

```
    - name: apply ingress acl to Ethernet 2/2
      nxos_acl_interface:
        name: border_inbound
        interface: Ethernet2/2
        direction: ingress
        state: present
        provider: "{{ cli }}"
```

The access list on VIRL NX-OSv is only supported on the management interface. You will see this warning: **Warning: ACL may not behave as expected since only management interface is supported** if you configure this ACL via the CLI. This warning is okay, as our purpose is only to demonstrate the configuration automation of the access list.

This might seem to be a lot of work for a single access list. For an experienced engineer, using Ansible to do this task will take longer than just logging in to the device and configuring the access list. However, remember that this playbook can be reused many times in the future, so it will save you time in the long run.

It is my experience that often, for a long access list, a few entries will be for one service, a few entries will be for another, and so on. The access lists tend to grow organically over time, and it becomes very hard to keep track of the origin and purpose of each entry. The fact that we can break them apart makes management of a long access list much simpler.

MAC access lists

In the case where you have an L2 environment or where you are using non-IP protocols on Ethernet interfaces, you can still use a MAC address access list to allow or deny hosts based on MAC addresses. The steps are similar to the IP access list but the match will be based on MAC addresses. Recall that for MAC addresses, or physical addresses, the first six hexadecimal symbols belong to an **Organizationally Unique Identifier** (**OUI**). So, we can use the same access list matching pattern to deny a certain group of hosts.

We are testing this on IOSv with the `ios_config` module. For older Ansible versions, the change will be pushed out every single time the playbook is executed. For newer Ansible versions, the control node will check for change first and only make changes when needed.

The host file and the top portion of the playbook are similar to the IP access list; the `tasks` portion is where the different modules and arguments are used:

```
<skip>
  tasks:
    - name: Deny Hosts with vendor id fa16.3e00.0000
      ios_config:
        lines:
          - access-list 700 deny fa16.3e00.0000 0000.00FF.FFFF
          - access-list 700 permit 0000.0000.0000 FFFF.FFFF.FFFF
        provider: "{{ cli }}"
    - name: Apply filter on bridge group 1
      ios_config:
        lines:
          - bridge-group 1
          - bridge-group 1 input-address-list 700
        parents:
          - interface GigabitEthernet0/1
        provider: "{{ cli }}"
```

As more virtual networks become popular, the L3 information sometimes becomes transparent to the underlying virtual links. In these scenarios, the MAC access list becomes a good option if you need to restrict access to those links.

The Syslog search

There are plenty of documented network security breaches that took place over an extended period of time. In these slow breaches, quite often, we saw signs and traces in logs indicating that there were suspicious activities. These can be found in both server and network device logs. The activities were not detected, not because there was a lack of information, but rather because there was too much information. The critical information that we were looking for is usually buried deep in a mountain of information that is hard to sort out.

Besides Syslog, **Uncomplicated Firewall** (**UFW**) is another great source of log information for servers. It is a frontend to iptables, which is a server firewall. UFW makes managing firewall rules very simple and logs a good amount of information. See the *Other tools* section for more information on UFW.

In this section, we will try to use Python to search through the Syslog text in order to detect the activities that we were looking for. Of course, the exact terms that we will search for depend on the device we are using. For example, Cisco provides a list of messages to look for in Syslog for any the access list violation logging. It is available at http://www.cisco.com/c/en/us/about/security-center/identify-incidents-via-syslog.html.

For more understanding of access control list logging, go to http://www.cisco.com/c/en/us/about/security-center/access-control-list-logging.html.

For our exercise, we will use a Nexus switch anonymized syslog file containing about 65,000 lines of log messages this file is included in the accommodated book GitHub repository for you:

```
$ wc -l sample_log_anonymized.log
65102 sample_log_anonymized.log
```

We have inserted some Syslog messages from the Cisco documentation (http://www.cisco.com/c/en/us/support/docs/switches/nexus-7000-series-switches/118907-configure-nx7k-00.html) as the log message that we should be looking for:

```
2014 Jun 29 19:20:57 Nexus-7000 %VSHD-5-VSHD_SYSLOG_CONFIG_I: Configured from vty by admin on console0
2014 Jun 29 19:21:18 Nexus-7000 %ACLLOG-5-ACLLOG_FLOW_INTERVAL: Src IP: 10.1 0.10.1,
 Dst IP: 172.16.10.10, Src Port: 0, Dst Port: 0, Src Intf: Ethernet4/1, Protocol: "ICMP"(1), Hit-count = 2589
```

```
2014 Jun 29 19:26:18 Nexus-7000 %ACLLOG-5-ACLLOG_FLOW_INTERVAL: Src IP:
10.1 0.10.1, Dst IP: 172.16.10.10, Src Port: 0, Dst Port: 0, Src Intf:
Ethernet4/1, Pro tocol: "ICMP"(1), Hit-count = 4561
```

We will be using simple examples with regular expressions. If you are already familiar with the regular expression in Python, feel free to skip the rest of the section.

Searching with the RE module

For our first search, we will simply use the regular expression module to look for the terms we are looking for. We will use a simple loop to the following:

```
#!/usr/bin/env python3

import re, datetime

startTime = datetime.datetime.now()

with open('sample_log_anonymized.log', 'r') as f:
    for line in f.readlines():
        if re.search('ACLLOG-5-ACLLOG_FLOW_INTERVAL', line):
            print(line)

endTime = datetime.datetime.now()
elapsedTime = endTime - startTime
print("Time Elapsed: " + str(elapsedTime))
```

The result took about 6/100 of a second to search through the log file:

```
$ python3 python_re_search_1.py
2014 Jun 29 19:21:18 Nexus-7000 %ACLLOG-5-ACLLOG_FLOW_INTERVAL: Src IP:
10.1 0.10.1,

2014 Jun 29 19:26:18 Nexus-7000 %ACLLOG-5-ACLLOG_FLOW_INTERVAL: Src IP:
10.1 0.10.1,

Time Elapsed: 0:00:00.065436
```

It is recommended to compile the search term for a more efficient search. It will not impact us much since the script is already pretty fast. In fact, the Python interpretative nature might actually make it slower. However, it will make a difference when we search through a larger text body, so let's make the change:

```
searchTerm = re.compile('ACLLOG-5-ACLLOG_FLOW_INTERVAL')

with open('sample_log_anonymized.log', 'r') as f:
```

```
        for line in f.readlines():
            if re.search(searchTerm, line):
                print(line)
```

The timing result is actually slower:

```
Time Elapsed: 0:00:00.081541
```

Let's expand the example a bit. Assuming we have several files and multiple terms to search through, we will copy the original file to a new file:

```
$ cp sample_log_anonymized.log sample_log_anonymized_1.log
```

We will also include searching for the `PAM: Authentication failure` term. We will add another loop to search both the files:

```
term1 = re.compile('ACLLOG-5-ACLLOG_FLOW_INTERVAL')
term2 = re.compile('PAM: Authentication failure')

fileList = ['sample_log_anonymized.log', 'sample_log_anonymized_1.log']

for log in fileList:
    with open(log, 'r') as f:
        for line in f.readlines():
            if re.search(term1, line) or re.search(term2, line):
                print(line)
```

We can now see the difference in performance by expanding our search terms and the number of messages:

```
$ python3 python_re_search_2.py
2016 Jun 5 16:49:33 NEXUS-A %DAEMON-3-SYSTEM_MSG: error: PAM:
Authentication failure for illegal user AAA from 172.16.20.170 - sshd[4425]

2016 Sep 14 22:52:26.210 NEXUS-A %DAEMON-3-SYSTEM_MSG: error: PAM:
Authentication failure for illegal user AAA from 172.16.20.170 - sshd[2811]

<skip>

2014 Jun 29 19:21:18 Nexus-7000 %ACLLOG-5-ACLLOG_FLOW_INTERVAL: Src IP:
10.1 0.10.1,

2014 Jun 29 19:26:18 Nexus-7000 %ACLLOG-5-ACLLOG_FLOW_INTERVAL: Src IP:
10.1 0.10.1,

<skip>

Time Elapsed: 0:00:00.330697
```

Of course, when it comes to performance tuning, it is a never-ending, impossible race to zero, and the performance sometimes depends on the hardware you are using. But the important point is to regularly perform audits of your log files using Python, so you can catch the early signals of any potential breach.

Other tools

There are other network security tools that we can use and automate with Python. Let's take a look at a few of them.

Private VLANs

Virtual Local Area Networks (VLANs) have been around for a long time. They are essentially a broadcast domain where all hosts can be connected to a single switch, but are partitioned out to different domains, so we can separate the hosts out according to which host can see others via broadcasts. Let's look at an mapped based on IP subnets. For example, in an enterprise building, I would likely see one IP subnet per physical floor: `192.168.1.0/24` for the first floor, `192.168.2.0/24` for the second floor, and so on. In this pattern, we use a 1/24 block for each floor. This gives a clear delineation of my physical network as well as my logical network. A host wanting to communicate beyond its own subnet will need to traverse through its layer 3 gateway, where I can use an access list to enforce security.

What happens when different departments reside on the same floor? Perhaps the finance and sales teams are both on the second floor, and I would not want the sales team's hosts in the same broadcast domain as the finance team's. I can break the subnet down further, but that might become tedious and break the standard subnet scheme that was previously set up. This is where a private VLAN can help.

The private VLAN essentially breaks up the existing VLAN into sub-VLANs. There are three categories within a private VLAN:

- **The Promiscuous (P) port**: This port is allowed to send and receive layer 2 frames from any other port on the VLAN; this usually belongs to the port connecting to the layer 3 router

- **The Isolated (I) port**: This port is only allowed to communicate with P ports, and they are typically connected to hosts when you do not want it to communicate with other hosts in the same VLAN
- **The Community (C) port**: This port is allowed to communicate with other C ports in the same community and P ports

We can again use Ansible or any of the other Python scripts introduced so far to accomplish this task. By now, we should have enough practice and confidence to implement this feature via automation, so I will not repeat the steps here. Being aware of the private VLAN feature would come in handy at times when you need to isolate ports even further in an L2 VLAN.

UFW with Python

We briefly mentioned UFW as the frontend for iptables on Ubuntu hosts. Here is a quick overview:

```
$ sudo apt-get install ufw
$ sudo ufw status
$ sudo ufw default outgoing
$ sudo ufw allow 22/tcp
$ sudo ufw allow www
$ sudo ufw default deny incoming
```

We can see the status of UFW:

```
$ sudo ufw status verbose
Status: active
Logging: on (low)
Default: deny (incoming), allow (outgoing), disabled (routed)
New profiles: skip

To                         Action      From
--                         ------      ----
22/tcp                     ALLOW IN    Anywhere
80/tcp                     ALLOW IN    Anywhere
22/tcp (v6)                ALLOW IN    Anywhere (v6)
80/tcp (v6)                ALLOW IN    Anywhere (v6)
```

[205]

As you can see, the advantage of UFW is a simple interface to construct otherwise complicated IP table rules. There are several Python-related tools we can use with UFW to make things even simpler:

- We can use the Ansible UFW module to streamline our operations. More information is available at `http://docs.ansible.com/ansible/ufw_module.html`. Because Ansible is written in Python, we can go further and examine what is inside the Python module source code. More information is available at `https://github.com/ansible/ansible/blob/devel/lib/ansible/modules/system/ufw.py`.
- There are Python wrapper modules around UFW as an API (visit `https://gitlab.com/dhj/easyufw`). This can make integration easier if you need to dynamically modify UFW rules based on certain events.
- UFW itself is written in Python. Therefore, you can use the existing Python knowledge if you ever need to extend the current command sets. More information is available at `https://launchpad.net/ufw`.

UFW proves to be a good tool to safeguard your network server.

Further reading

Python is a very common language used in many of the security-related fields. A few of the books I would recommend are listed as follows:

- **Violent Python**: A cookbook for hackers, forensic analysts, penetration testers, and security engineers by T.J. O'Connor (ISBN-10: 1597499579)
- **Black Hat Python**: Python programming for hackers and pentesters by Justin Seitz (ISBN-10: 1593275900)

I have personally used Python extensively in our research work on **Distributed Denial of Service (DDoS)** at A10 Networks. If you are interested in learning more, the guide can be downloaded for free at `https://www.a10networks.com/resources/ebooks/distributed-denial-service-ddos`.

Summary

In this chapter, we looked at network security with Python. We used the Cisco VIRL tool to set up our lab with both hosts and network devices, consisting of NX-OSv and IOSv types. We took a tour around Scapy, which allows us to construct packets from the ground up. Scapy can be used in the interactive mode for quick testing. Once completed in interactive mode, we can put the steps into a file for more scalable testing. It can be used to perform various network penetration testing for known vulnerabilities.

We also looked at how we can use both an IP access list as well as a MAC access list to protect our network. They are usually the first line of defense in our network protection. Using Ansible, we are able to deploy access lists consistently and quickly to multiple devices.

Syslog and other log files contain useful information that we should regularly comb through to detect any early signs of a breach. Using Python regular expressions, we can systematically search for known log entries that can point us to security events that require our attention. Besides the tools we have discussed, private VLAN and UFW are among some other useful tools that we can use for more security protection.

In `Chapter 7`, *Network Monitoring with Python – Part 1*, we will look at how to use Python for network monitoring. Monitoring allows us to know what is happening in our network and the state of the network.

7
Network Monitoring with Python – Part 1

Imagine you get a call at 2:00 a.m. in the morning. The person on the other end says: "Hi, we are facing a difficult issue that is impacting production services. We suspect it might be network-related. Can you check this for us? For this type of urgent, open-ended question, what would be the first thing you do?" Most of the time, the thing that comes to mind would be: What changed between the time when the network was working until something went wrong? Chances are you would check your monitoring tool and see if any of the key metrics changed in the last few hours. Better yet is if you have received any monitoring alerts from any metric baseline deviation.

Throughout this book, we have been discussing various ways to systematically make predictable changes to our network, with the goal of keeping the network running as smoothly as possible. However, networks are not static – far from it – they are probably one of the most fluid parts of the entire infrastructure. By definition, a network connects different parts of the infrastructure together, constantly passing traffic back and forth. There are lots of moving parts that can cause your network to stop working as expected: hardware failures, software with bugs, human mistakes despite their best intentions, and many more. It is not a question of whether things would go wrong, but rather a question of when and what went wrong when it happens. We need ways to monitor our network to make sure it works as expected and hopefully be notified when it does not.

In upcoming two chapters, we will look at various ways to perform network monitoring tasks. Many of the tools we have looked at thus far can be tied together or directly managed by Python. Like many tools we have looked at, network monitoring has to do with two parts. First, we need to know what information the equipment is capable of transmitting. Second, we need to identify what useful information we can interpret from them.

We will look at a few tools that allow us to monitor the network effectively:

- The **Simple Network Management Protocol (SNMP)**
- Matplotlib and Pygal visualization
- MRTG and Cacti

This list is not exhaustive, and there is certainly no lack of commercial vendors in the network monitoring space. The basics of network monitoring that we will look at, however, carry well for both open source and commercial tools.

Lab setup

The lab for this chapter is similar to the one in `Chapter 6`, *Network Security with Python*, but with this difference: both of the network devices are IOSv devices. Here's an illustration of this:

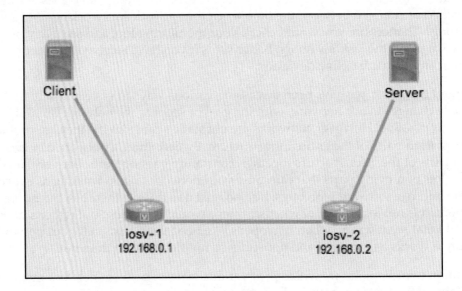

The two Ubuntu hosts will be used to generate traffic across the network so that we can look at some non-zero counters.

SNMP

SNMP is a standardized protocol used to collect and manage devices. Although the standard allows you to use SNMP for device management, in my experience, most network administrators prefer to keep SNMP as an information collection mechanism only. Since SNMP operates on UDP, which is connectionless, and considering the relatively weak security mechanism in versions 1 and 2, making device changes via SNMP tends to make network operators a bit uneasy. SNMP version 3 has added cryptographic security and new concepts and terminologies to the protocol, but the way the technology is adapted varies among network device vendors.

SNMP is widely used in network monitoring and has been around since 1988 as part of RFC 1065. The operations are straightforward, with the network manager sending GET and SET requests toward the device and the device with the SNMP agent responding with the information per request. The most widely adopted standard is SNMPv2c, which is defined in RFC 1901 – RFC 1908. It uses a simple community-based security scheme for security. It has also introduced new features, such as the ability to get bulk information. The following diagram displays the high-level operation for SNMP:

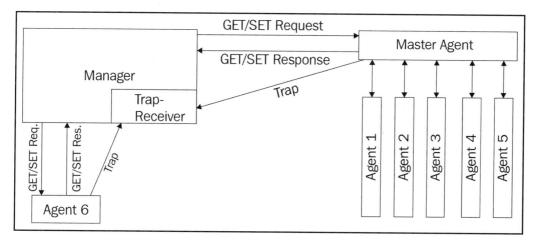

SNMP operations

The information residing in the device is structured in the **Management Information Base (MIB)**. The MIB uses a hierarchical namespace containing an **Object Identifier (OID)**, which represents the information that can be read and fed back to the requester. When we talk about using SNMP to query device information, we are really talking about using the management station to query the specific OID that represents the information we are after. There is a common OID structure, such as systems and interfaces OID, that is shared among vendors. Besides common OID, each vendor can also supply an enterprise-level OID that is specific to them.

As an operator, you are required to put some effort into consolidating information into an OID structure in your environment to retrieve useful information. This can sometimes be a tedious process of finding one OID at a time. For example, you might be making a request to a device OID and receive a value of 10,000. What is that value? Is that interface traffic? Is it in bytes or bits? Or maybe it is a number of packets? How do we know? We will need to consult either the standard or the vendor documentation to find out. There are tools that help with this process, such as a MIB browser that can provide more metadata to the value. But, at least in my experience, constructing an SNMP-based monitoring tool for your network can sometimes feel like a cat-and-mouse game of trying to find that one missing value.

Some of the main points to take away from the operation are as follows:

- The implementation relies heavily on the amount of information the device agent can provide. This, in turn, relies on how the vendor treats SNMP: as a core feature or an added feature.
- SNMP agents generally require CPU cycles from the control plane to return a value. Not only is this inefficient for devices with, say, large BGP tables, it is also not feasible to use SNMP to query the data at small intervals.
- The user needs to know the OID in order to query the data.

Since SNMP has been around for a while, my assumption is that you have some experience with it already. Let's jump directly into package installation and our first SNMP example.

Setup

First, let's make sure that we have the SNMP managing device and agent work in our setup. The SNMP bundle can be installed on either the hosts (client or server) in our lab or the managing device on the management network. As long as the SNMP manager has IP reachability to the device and the managed device allows the inbound connection, SNMP should work. In production, you should only install the software on the management host and only allow SNMP traffic in the control plane.

In this lab, we have installed SNMP on both the Ubuntu host on the management network and the client host in the lab to test security:

```
$ sudo apt-get install snmp
```

The next step would be to turn on and configure the SNMP options on the network devices, `iosv-1` and `iosv-2`. There are many optional parameters you can configure on the network device, such as contact, location, chassis ID, and SNMP packet size. The options are device-specific and you should check the documentation on your device. For IOSv devices, we will configure an access list to limit only the desired host for querying the device as well as tying the access list with the SNMP community string. In our case, we will use the word `secret` as the read-only community string and `permit_snmp` as the access list name:

```
!
ip access-list standard permit_snmp
 permit 172.16.1.173 log
 deny any log
!
!
snmp-server community secret RO permit_snmp
!
```

The SNMP community string is acting as a shared password between the manager and the agent; therefore, it needs to be included any time you want to query the device.

As mentioned earlier in this chapter, finding the right OID is oftentimes half of the battle when working with SNMP. We can use tools such as the Cisco IOS MIB locator (http://tools.cisco.com/ITDIT/MIBS/servlet/index) for finding specific OIDs to query. Alternatively, we can just start walking through the SNMP tree, starting from the top of Cisco's enterprise tree at .1.3.6.1.4.1.9. We will perform the walk to make sure that the SNMP agent and the access-list are working:

```
$ snmpwalk -v2c -c secret 172.16.1.189 .1.3.6.1.4.1.9
iso.3.6.1.4.1.9.2.1.1.0 = STRING: "
Bootstrap program is IOSv
"
iso.3.6.1.4.1.9.2.1.2.0 = STRING: "reload"
iso.3.6.1.4.1.9.2.1.3.0 = STRING: "iosv-1"
iso.3.6.1.4.1.9.2.1.4.0 = STRING: "virl.info"
...
```

We can be more specific about the OID we need to query as well:

```
$ snmpwalk -v2c -c secret 172.16.1.189 .1.3.6.1.4.1.9.2.1.61.0
iso.3.6.1.4.1.9.2.1.61.0 = STRING: "cisco Systems, Inc.
170 West Tasman Dr.
San Jose, CA 95134-1706
U.S.A.
Ph +1-408-526-4000
Customer service 1-800-553-6387 or +1-408-526-7208
24HR Emergency 1-800-553-2447 or +1-408-526-7209
Email Address tac@cisco.com
World Wide Web http://www.cisco.com"
```

As a matter of demonstration, what if we type in the wrong value by 1 digit from 0 to 1 at the end of the last OID? This is what we would see:

```
$ snmpwalk -v2c -c secret 172.16.1.189 .1.3.6.1.4.1.9.2.1.61.1
iso.3.6.1.4.1.9.2.1.61.1 = No Such Instance currently exists at this OID
```

Unlike API calls, there are no useful error codes nor messages; it simply stated that the OID does not exist. This can be pretty frustrating at times.

The last thing to check would be the access list we configured will deny unwanted SNMP queries. Because we had the `log` keyword for both the permit and deny entries in the access-list, only `172.16.1.173` is permitted to query the devices:

```
*Mar 3 20:30:32.179: %SEC-6-IPACCESSLOGNP: list permit_snmp permitted 0 172.16.1.173 -> 0.0.0.0, 1 packet
*Mar 3 20:30:33.991: %SEC-6-IPACCESSLOGNP: list permit_snmp denied 0 172.16.1.187 -> 0.0.0.0, 1 packet
```

As you can see, the biggest challenge in setting up SNMP is to find the right OID. Some of the OIDs are defined in standardized MIB-2; others are under the enterprise portion of the tree. Vendor documentation is the best bet, though. There are a number of tools that can help, such as a MIB browser; you can add MIBs (again, provided by the vendors) to the browser and see the description of the enterprise-based OIDs. A tool such as Cisco's SNMP Object Navigator (`http://snmp.cloudapps.cisco.com/Support/SNMP/do/BrowseOID.do?local=en`) proves to be very valuable when you need to find the correct OID of the object you are looking for.

Chapter 7

PySNMP

PySNMP is a cross-platform, pure Python SNMP engine implementation developed by Ilya Etingof (https://github.com/etingof). It abstracts a lot of SNMP details for you, as great libraries do, and supports both Python 2 and Python 3.

PySNMP requires the PyASN1 package. The following is taken from to Wikipedia:

> "ASN.1 is a standard and notation that describes rules and structures for representing, encoding, transmitting, and decoding data in telecommunication and computer networking."

PyASN1 conveniently provides a Python wrapper around ASN.1. Let's install the package first:

```
cd /tmp
git clone https://github.com/etingof/pyasn1.git
cd pyasn1/
git checkout 0.2.3
sudo python3 setup.py install
```

Next, install the PySNMP package:

```
git clone https://github.com/etingof/pysnmp
cd pysnmp/
git checkout v4.3.10
sudo python3 setup.py install
```

We are using an older version of PySNMP due to the fact that pysnmp.entity.rfc3413.oneliner was removed starting with version 5.0.0 (https://github.com/etingof/pysnmp/blob/a93241007b970c458a0233c16ae2ef82dc107290/CHANGES.txt). If you use pip to install the packages, the examples will likely break.

[215]

Network Monitoring with Python – Part 1

Let's look at how to use PySNMP to query the same Cisco contact information we used in the previous example. The steps we will take are slightly modified versions from the PySNMP example at http://pysnmp.sourceforge.net/faq/response-values-mib-resolution.html. We will import the necessary module and create a CommandGenerator object first:

```
>>> from pysnmp.entity.rfc3413.oneliner import cmdgen
>>> cmdGen = cmdgen.CommandGenerator()
>>> cisco_contact_info_oid = "1.3.6.1.4.1.9.2.1.61.0"
```

We can perform SNMP using the getCmd method. The result is unpacked into various variables; of these, we care most about varBinds, which contains the query result:

```
>>> errorIndication, errorStatus, errorIndex, varBinds = cmdGen.getCmd(
...     cmdgen.CommunityData('secret'),
...     cmdgen.UdpTransportTarget(('172.16.1.189', 161)),
...     cisco_contact_info_oid
... )
>>> for name, val in varBinds:
...     print('%s = %s' % (name.prettyPrint(), str(val)))
...
SNMPv2-SMI::enterprises.9.2.1.61.0 = cisco Systems, Inc.
170 West Tasman Dr.
San Jose, CA 95134-1706
U.S.A.
Ph +1-408-526-4000
Customer service 1-800-553-6387 or +1-408-526-7208
24HR Emergency 1-800-553-2447 or +1-408-526-7209
Email Address tac@cisco.com
World Wide Web http://www.cisco.com
>>>
```

Note that the response values are PyASN1 objects. The prettyPrint() method will convert some of these values into a human-readable format, but since the result in our case was not converted, we will convert it into a string manually.

We can write a script based on the preceding interactive example. We will name it pysnmp_1.py with error checking. We can also include multiple OIDs in the getCmd() method:

```
#!/usr/bin/env/python3

from pysnmp.entity.rfc3413.oneliner import cmdgen

cmdGen = cmdgen.CommandGenerator()

system_up_time_oid = "1.3.6.1.2.1.1.3.0"
```

```
cisco_contact_info_oid = "1.3.6.1.4.1.9.2.1.61.0"

errorIndication, errorStatus, errorIndex, varBinds = cmdGen.getCmd(
    cmdgen.CommunityData('secret'),
    cmdgen.UdpTransportTarget(('172.16.1.189', 161)),
    system_up_time_oid,
    cisco_contact_info_oid
)

# Check for errors and print out results
if errorIndication:
    print(errorIndication)
else:
    if errorStatus:
        print('%s at %s' % (
            errorStatus.prettyPrint(),
            errorIndex and varBinds[int(errorIndex)-1] or '?'
            )
        )
    else:
        for name, val in varBinds:
            print('%s = %s' % (name.prettyPrint(), str(val)))
```

The result will be unpacked and list out the values of the two OIDs:

```
$ python3 pysnmp_1.py
SNMPv2-MIB::sysUpTime.0 = 660959
SNMPv2-SMI::enterprises.9.2.1.61.0 = cisco Systems, Inc.
170 West Tasman Dr.
San Jose, CA 95134-1706
U.S.A.
Ph +1-408-526-4000
Customer service 1-800-553-6387 or +1-408-526-7208
24HR Emergency 1-800-553-2447 or +1-408-526-7209
Email Address tac@cisco.com
World Wide Web http://www.cisco.com
```

In the following example, we will persist the values we received from the queries so that we can perform other functions, such as visualization, with the data. For our example, we will use `ifEntry` within the MIB-2 tree for interface-related values to be graphed. You can find a number of resources that map out the `ifEntry` tree; here is a screenshot of the Cisco SNMP Object Navigator site that we accessed previously for `ifEntry`:

```
OID Tree
You are currently viewing your object with  2   levels of hierarchy above your object.

. iso (1) . org (3) . dod (6) . internet (1) . mgmt (2) . mib-2 (1)
  |
  - -- interfaces (2)
      |
      | -- ifNumber (1)
      |
      - -- ifTable (2)
          |
          - -- ifEntry (1) object Details
              |
              | -- ifIndex (1)
              |
              | -- ifDescr (2)
              |
              | -- ifType (3)
              |
              | -- ifMtu (4)
              |
              | -- ifSpeed (5)
              |
              | -- ifPhysAddress (6)
              |
              | -- ifAdminStatus (7)
              |
              | -- ifOperStatus (8)
              |
              | -- ifLastChange (9)
              |
              | -- ifInOctets (10)
              |
              | -- ifInUcastPkts (11)
              |
              | -- ifInNUcastPkts (12)
              |
              | -- ifInDiscards (13)
              |
              | -- ifInErrors (14)
              |
              | -- ifInUnknownProtos (15)
              |
              | -- ifOutOctets (16)
              |
              | -- ifOutUcastPkts (17)
              |
              | -- ifOutNUcastPkts (18)
              |
              | -- ifOutDiscards (19)
              |
              | -- ifOutErrors (20)
              |
              | -- ifOutQLen (21)
              |
              | -- ifSpecific (22)
```

SNMP ifEntry OID tree

A quick test will illustrate the OID mapping of the interfaces on the device:

```
$ snmpwalk -v2c -c secret 172.16.1.189 .1.3.6.1.2.1.2.2.1.2
iso.3.6.1.2.1.2.2.1.2.1 = STRING: "GigabitEthernet0/0"
iso.3.6.1.2.1.2.2.1.2.2 = STRING: "GigabitEthernet0/1"
iso.3.6.1.2.1.2.2.1.2.3 = STRING: "GigabitEthernet0/2"
iso.3.6.1.2.1.2.2.1.2.4 = STRING: "Null0"
iso.3.6.1.2.1.2.2.1.2.5 = STRING: "Loopback0"
```

From the documentation, we can map the values of `ifInOctets(10)`, `ifInUcastPkts(11)`, `ifOutOctets(16)`, and `ifOutUcastPkts(17)` into their respective OID values. From a quick check on the CLI and MIB documentation, we can see that the value of the `GigabitEthernet0/0` packets output maps to OID `1.3.6.1.2.1.2.2.1.17.1`. We will follow the rest of the same process to map out the rest of the OIDs for the interface statistics. When checking between CLI and SNMP, keep in mind that the values should be close but not exactly the same since there might be some traffic on the wire between the time of CLI output and the SNMP query time:

```
# Command Line Output
iosv-1#sh int gig 0/0 | i packets
  5 minute input rate 0 bits/sec, 0 packets/sec
  5 minute output rate 0 bits/sec, 0 packets/sec
     38532 packets input, 3635282 bytes, 0 no buffer
     53965 packets output, 4723884 bytes, 0 underruns

# SNMP Output
$ snmpwalk -v2c -c secret 172.16.1.189 .1.3.6.1.2.1.2.2.1.17.1
iso.3.6.1.2.1.2.2.1.17.1 = Counter32: 54070
```

If we are in a production environment, we will likely write the results into a database. But since this is just an example, we will write the query values to a flat file. We will write the `pysnmp_3.py` script for information query and write the results to the file. In the script, we have defined various OIDs that we need to query:

```
# Hostname OID
system_name = '1.3.6.1.2.1.1.5.0'

# Interface OID
gig0_0_in_oct = '1.3.6.1.2.1.2.2.1.10.1'
gig0_0_in_uPackets = '1.3.6.1.2.1.2.2.1.11.1'
gig0_0_out_oct = '1.3.6.1.2.1.2.2.1.16.1'
gig0_0_out_uPackets = '1.3.6.1.2.1.2.2.1.17.1'
```

The values were consumed in the `snmp_query()` function, with the `host`, `community`, and `oid` as input:

```python
def snmp_query(host, community, oid):
    errorIndication, errorStatus, errorIndex, varBinds = cmdGen.getCmd(
    cmdgen.CommunityData(community),
    cmdgen.UdpTransportTarget((host, 161)),
    oid
    )
```

All of the values are put in a dictionary with various keys and written to a file called `results.txt`:

```python
result = {}
result['Time'] = datetime.datetime.utcnow().isoformat()
result['hostname'] = snmp_query(host, community, system_name)
result['Gig0-0_In_Octet'] = snmp_query(host, community, gig0_0_in_oct)
result['Gig0-0_In_uPackets'] = snmp_query(host, community, gig0_0_in_uPackets)
result['Gig0-0_Out_Octet'] = snmp_query(host, community, gig0_0_out_oct)
result['Gig0-0_Out_uPackets'] = snmp_query(host, community, gig0_0_out_uPackets)

with open('/home/echou/Master_Python_Networking/Chapter7/results.txt', 'a') as f:
    f.write(str(result))
    f.write('n')
```

The outcome will be a file with results showing the interface packets represented at the time of the query:

```
# Sample output
$ cat results.txt
{'Gig0-0_In_Octet': '3990616', 'Gig0-0_Out_uPackets': '60077',
'Gig0-0_In_uPackets': '42229', 'Gig0-0_Out_Octet': '5228254', 'Time':
'2017-03-06T02:34:02.146245', 'hostname': 'iosv-1.virl.info'}
{'Gig0-0_Out_uPackets': '60095', 'hostname': 'iosv-1.virl.info',
'Gig0-0_Out_Octet': '5229721', 'Time': '2017-03-06T02:35:02.072340',
'Gig0-0_In_Octet': '3991754', 'Gig0-0_In_uPackets': '42242'}
{'hostname': 'iosv-1.virl.info', 'Gig0-0_Out_Octet': '5231484',
'Gig0-0_In_Octet': '3993129', 'Time': '2017-03-06T02:36:02.753134',
'Gig0-0_In_uPackets': '42257', 'Gig0-0_Out_uPackets': '60116'}
{'Gig0-0_In_Octet': '3994504', 'Time': '2017-03-06T02:37:02.146894',
'Gig0-0_In_uPackets': '42272', 'Gig0-0_Out_uPackets': '60136',
'Gig0-0_Out_Octet': '5233187', 'hostname': 'iosv-1.virl.info'}
{'Gig0-0_In_uPackets': '42284', 'Time': '2017-03-06T02:38:01.915432',
```

```
'Gig0-0_In_Octet': '3995585', 'Gig0-0_Out_Octet': '5234656',
'Gig0-0_Out_uPackets': '60154', 'hostname': 'iosv-1.virl.info'}
...
```

We can make this script executable and schedule a `cron` job to be executed every five minutes:

```
$ chmod +x pysnmp_3.py

# Crontab configuration
*/5 * * * * /home/echou/Master_Python_Networking/Chapter7/pysnmp_3.py
```

As mentioned previously, in a production environment, we would put the information in a database. For a SQL database, you can use a unique ID as the primary key. In a NoSQL database, we might use time as the primary index (or key) because it is always unique, followed by various key-value pairs.

We will wait for the script to be executed a few times for the values to be populated. If you are the impatient type, you can shorten the `cron` job interval to be one minute. After you see enough values in the `results.txt` file to make an interesting graph, we can move on to the next section to see how we can use Python to visualize the data.

Python for data visualization

We gather network data for the purpose of gaining insight into our network. One of the best ways to know what the data means is to visualize it with graphs. This is true for almost all data, but especially true for time series data in the context of network monitoring. How much data was transmitted over the network in the last week? What is the percentage of the TCP protocol among all of the traffic? These are values we can glean from using data-gathering mechanisms, such as SNMP, and we can produce visualization graphs with some of the popular Python libraries.

In this section, we will use the data we collected from the last section using SNMP and use two popular Python libraries, Matplotlib and Pygal, to graph them.

Matplotlib

Matplotlib (http://matplotlib.org/) is a Python 2D plotting library for the Python language and its NumPy mathematical extension. It can produce publication-quality figures, such as plots, histograms, and bar graphs, with a few lines of code.

 NumPy is an extension of the Python programming language. It is open source and widely used in various data science projects. You can learn more about it at https://en.wikipedia.org/wiki/NumPy.

Installation

The installation can be done using the Linux package management system, depending on your distribution:

```
$ sudo apt-get install python-matplotlib # for Python2
$ sudo apt-get install python3-matplotlib
```

Matplotlib – the first example

For the following examples, the output figures are displayed as the standard output by default. During development, it is often easier to try out the code initially and produce the graph on the standard output first before finalizing the code with a script. If you have been following along with this book via a virtual machine, it is recommended that you use the VM window instead of SSH so that you can see the graphs. If you do not have access to the standard output, you can save the figure and view it after you download it (as you will see soon). Note that you will need to set the $DISPLAY variable in some of the graphs that we will produce in this section.

The following is a screenshot of the Ubuntu desktop used in this chapter's visualization example. As soon as the plt.show() command is issued in the Terminal window, Figure 1 will appear on the screen. When you close the figure, you will return to the Python shell:

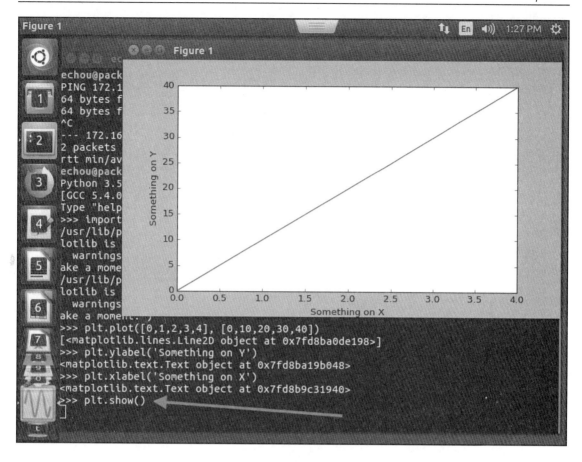

Matplotlib visualization with the Ubuntu desktop

Let's look at the line graph first. A line graph simply gives two lists of numbers that correspond to the *x* axis and *y* axis values:

```
>>> import matplotlib.pyplot as plt
>>> plt.plot([0,1,2,3,4], [0,10,20,30,40])
[<matplotlib.lines.Line2D object at 0x7f932510df98>]
>>> plt.ylabel('Something on Y')
<matplotlib.text.Text object at 0x7f93251546a0>
>>> plt.xlabel('Something on X')
<matplotlib.text.Text object at 0x7f9325fdb9e8>
>>> plt.show()
```

The graph will show up as a line graph:

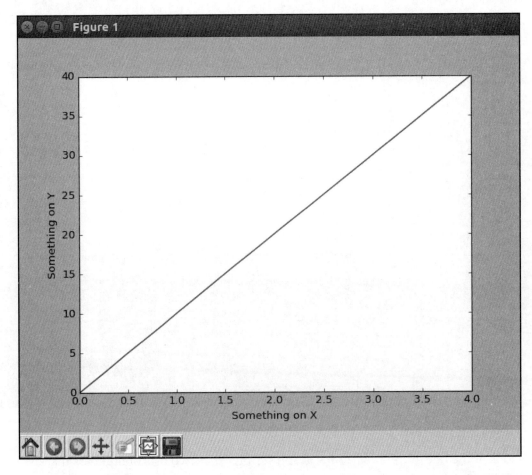

Matplotlib line graph

Alternatively, if you do not have access to standard output or have saved the figure first, you can use the `savefig()` method:

```
>>> plt.savefig('figure1.png')
or
>>> plt.savefig('figure1.pdf')
```

With this basic knowledge of graphing plots, we can now graph the results we receive from SNMP queries.

Matplotlib for SNMP results

In our first Matplotlib example, namely `matplotlib_1.py`, we will import the *dates* module besides `pyplot`. We will use the `matplotlib.dates` module instead of the Python standard library `dates` module. Unlike the Python `dates` module, the `mapplotlib.dates` library will convert the date value internally into the float type, which is required by Matplotlib:

```
import matplotlib.pyplot as plt
import matplotlib.dates as dates
```

> Matplotlib provides sophisticated date plotting capabilities; you can find more information on this at `http://matplotlib.org/api/dates_api.html`.

In the script, we will create two empty lists, each representing the *x*-axis and *y*-axis values. Note that, on line 12, we used the built-in `eval()` Python function to read the input as a dictionary instead of a default string:

```
x_time = []
y_value = []

with open('results.txt', 'r') as f:
    for line in f.readlines():
        line = eval(line)
        x_time.append(dates.datestr2num(line['Time']))
        y_value.append(line['Gig0-0_Out_uPackets'])
```

In order to read the *x*-axis value back in a human-readable date format, we will need to use the `plot_date()` function instead of `plot()`. We will also tweak the size of the figure a bit as well as rotate the value on the *x*-axis so that we can read the value in full:

```
plt.subplots_adjust(bottom=0.3)
plt.xticks(rotation=80)

plt.plot_date(x_time, y_value)
plt.title('Router1 G0/0')
plt.xlabel('Time in UTC')
plt.ylabel('Output Unicast Packets')
plt.savefig('matplotlib_1_result.png')
plt.show()
```

Network Monitoring with Python – Part 1

The final result will display the **Router1 Gig0/0** and **Output Unicast Packet**, as follows:

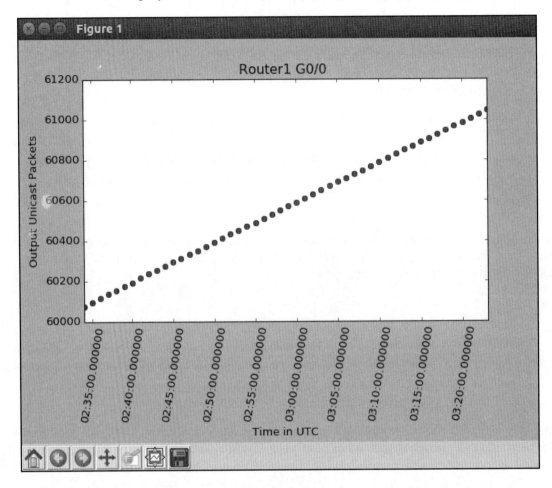

Router1 Matplotlib graph

Note that if you prefer a straight line instead of dots, you can use the third optional parameter in the `plot_date()` function:

```
plt.plot_date(x_time, y_value, "-")
```

Chapter 7

We can repeat the steps for the rest of the values for output octets, input unicast packets, and input as individual graphs. However, in our coming example, that is, `matplotlib_2.py`, we will show you how to graph multiple values against the same time range, as well as additional Matplotlib options.

In this case, we will create additional lists and populate the values accordingly:

```
x_time = []
out_octets = []
out_packets = []
in_octets = []
in_packets = []

with open('results.txt', 'r') as f:
    for line in f.readlines():
...
        out_packets.append(line['Gig0-0_Out_uPackets'])
        out_octets.append(line['Gig0-0_Out_Octet'])
        in_packets.append(line['Gig0-0_In_uPackets'])
        in_octets.append(line['Gig0-0_In_Octet'])
```

Since we have identical *x*-axis values, we can just add the different *y*-axis values to the same graph:

```
# Use plot_date to display x-axis back in date format
plt.plot_date(x_time, out_packets, '-', label='Out Packets')
plt.plot_date(x_time, out_octets, '-', label='Out Octets')
plt.plot_date(x_time, in_packets, '-', label='In Packets')
plt.plot_date(x_time, in_octets, '-', label='In Octets')
```

Also, add grid and legend to the graph:

```
plt.legend(loc='upper left')
plt.grid(True)
```

Network Monitoring with Python – Part 1

The final result will combine all of the values in a single graph. Note that some of the values in the upper-left corner are blocked by the legend. You can resize the figure and/or use the pan/zoom option to move around the graph in order to see the value:

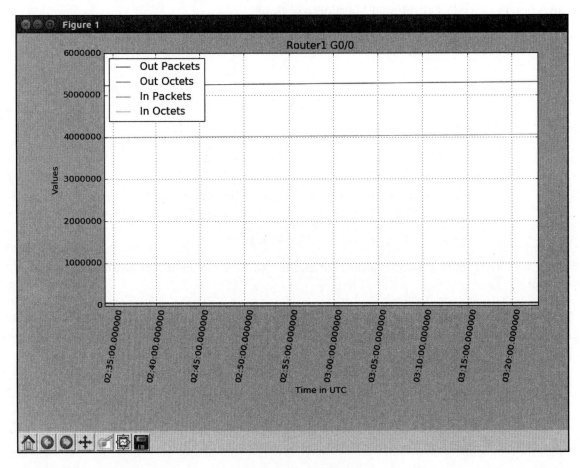

Router 1 – Matplotlib multiline graph

There are many more graphing options available in Matplotlib; we are certainly not limited to plot graphs. For example, we can use the following mock data to graph the percentage of different traffic types that we can see on the wire:

```
#!/usr/bin/env python3
# Example from
http://matplotlib.org/2.0.0/examples/pie_and_polar_charts/pie_demo_features.html
import matplotlib.pyplot as plt
```

Chapter 7

```
# Pie chart, where the slices will be ordered and plotted counter-
clockwise:
labels = 'TCP', 'UDP', 'ICMP', 'Others'
sizes = [15, 30, 45, 10]
explode = (0, 0.1, 0, 0) # Make UDP stand out

fig1, ax1 = plt.subplots()
ax1.pie(sizes, explode=explode, labels=labels, autopct='%1.1f%%',
 shadow=True, startangle=90)
ax1.axis('equal') # Equal aspect ratio ensures that pie is drawn as a
circle.

plt.show()
```

The preceding code leads to this pie chart from `plt.show()`:

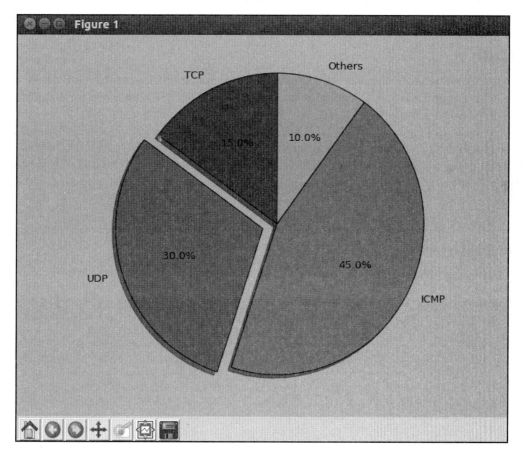

Matplotlib pie chart

Additional Matplotlib resources

Matplotlib is one of the best Python plotting libraries that is able to produce publication-quality figures. Like Python, its aim is to make complex tasks simple. With over 7,550 stars (and counting) on GitHub, it is also one of the most popular open source projects. Its popularity directly translates into faster bug fixes, a friendly user community, and general usability. It takes a bit of time to learn the package, but it is well worth the effort.

In this section, we barely scratched the surface of Matplotlib. You'll find additional resources at `http://matplotlib.org/2.0.0/index.html` (the Matplotlib project page) and `https://github.com/matplotlib/matplotlib` (the Matplotlib GitHub repository).

In the coming section, we will take a look at another popular Python graph library: **Pygal**.

Pygal

Pygal (`http://www.pygal.org/`) is a dynamic SVG charting library written in Python. The biggest advantage of Pygal, in my opinion, is that it produces **Scalable Vector Graphics** (**SVG**) format graphs easily and natively. There are many advantages of SVG over other graph formats, but two of the main advantages are that it is web browser-friendly and it provides scalability without sacrificing image quality. In other words, you can display the resulting image in any modern web browser and zoom in and out of the image without losing the details of the graph. Did I mention that we can do this in a few lines of Python code? How cool is that?

Installation

The installation is done via `pip`:

```
$ sudo pip install pygal #Python 2
$ sudo pip3 install pygal
```

Pygal – the first example

Let's look at the line chart example demonstrated on Pygal's documentation, available at http://pygal.org/en/stable/documentation/types/line.html:

```
>>> import pygal
>>> line_chart = pygal.Line()
>>> line_chart.title = 'Browser usage evolution (in %)'
>>> line_chart.x_labels = map(str, range(2002, 2013))
>>> line_chart.add('Firefox', [None, None, 0, 16.6, 25, 31, 36.4, 45.5, 46.3, 42.8, 37.1])
<pygal.graph.line.Line object at 0x7fa0bb009c50>
>>> line_chart.add('Chrome', [None, None, None, None, None, None, 0, 3.9, 10.8, 23.8, 35.3])
<pygal.graph.line.Line object at 0x7fa0bb009c50>
>>> line_chart.add('IE', [85.8, 84.6, 84.7, 74.5, 66, 58.6, 54.7, 44.8, 36.2, 26.6, 20.1])
<pygal.graph.line.Line object at 0x7fa0bb009c50>
>>> line_chart.add('Others', [14.2, 15.4, 15.3, 8.9, 9, 10.4, 8.9, 5.8, 6.7, 6.8, 7.5])
<pygal.graph.line.Line object at 0x7fa0bb009c50>
>>> line_chart.render_to_file('pygal_example_1.svg')
```

In this example, we created a line object with the x_labels automatically rendered as strings for 11 units. Each of the objects can be added with the label and the value in a list format, such as Firefox, Chrome, and IE.

Here's the resulting graph, as viewed in Firefox:

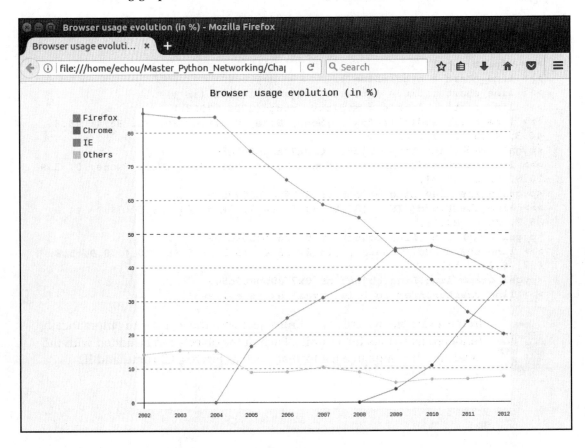

Pygal sample graph

Now that we can see the general usage of Pygal, we can use the same method to graph the SNMP results we have in hand. We will do this in the coming section.

Pygal for SNMP results

For the Pygal line graph, we can largely follow the same pattern as our Matplotlib example, where we create lists of values by reading the file. We no longer need to convert the *x*-axis value into an internal float, as we did for Matplotlib; however, we do need to convert the numbers in each of the values we would have received in the float:

```
#!/usr/bin/env python3

import pygal

x_time = []
out_octets = []
out_packets = []
in_octets = []
in_packets = []

with open('results.txt', 'r') as f:
    for line in f.readlines():
        line = eval(line)
        x_time.append(line['Time'])
        out_packets.append(float(line['Gig0-0_Out_uPackets']))
        out_octets.append(float(line['Gig0-0_Out_Octet']))
        in_packets.append(float(line['Gig0-0_In_uPackets']))
        in_octets.append(float(line['Gig0-0_In_Octet']))
```

We can use the same mechanism that we saw to construct the line graph:

```
line_chart = pygal.Line()
line_chart.title = "Router 1 Gig0/0"
line_chart.x_labels = x_time
line_chart.add('out_octets', out_octets)
line_chart.add('out_packets', out_packets)
line_chart.add('in_octets', in_octets)
line_chart.add('in_packets', in_packets)
line_chart.render_to_file('pygal_example_2.svg')
```

The outcome is similar to what we have already seen, but the graph is now in an SVG format that can be easily displayed on a web page. It can be viewed from a modern web browser:

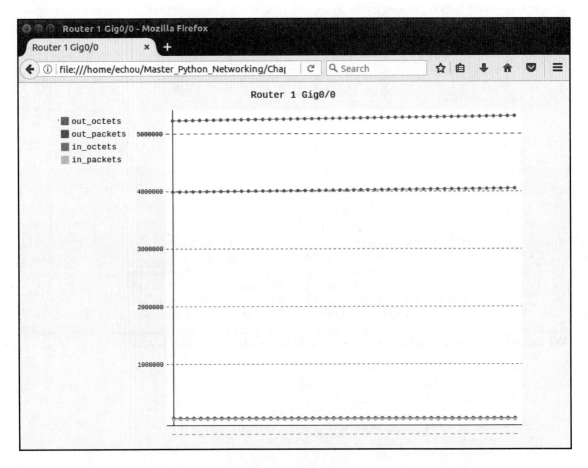

Router 1—Pygal multiline graph

Just like Matplotlib, Pygal provides many more options for graphs. For example, to graph the pie chart we saw previously in Pygal, we can use the `pygal.Pie()` object:

```
#!/usr/bin/env python3

import pygal

line_chart = pygal.Pie()
line_chart.title = "Protocol Breakdown"
line_chart.add('TCP', 15)
```

```
    line_chart.add('UDP', 30)
    line_chart.add('ICMP', 45)
    line_chart.add('Others', 10)
    line_chart.render_to_file('pygal_example_3.svg')
```

The resulting SVG file would be similar to the PNG generated by Matplotlib:

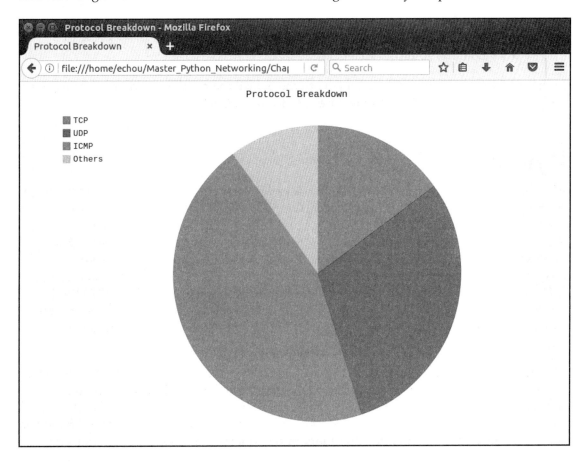

Pygal pie chart

Additional Pygal resources

Pygal provides many more customizable features and graphing capabilities for the data you collect from basic network monitoring tools such as SNMP. We demonstrated a simple line graph and pie graphs in this section. You can find more information about the project here:

- **Pygal documentation**: http://www.pygal.org/en/stable/index.html
- **Pygal GitHub project page**: https://github.com/Kozea/pygal

In the coming section, we will continue with the SNMP theme of network monitoring but with a fully featured network monitoring system called **Cacti**.

Python for Cacti

In my early days working as a junior network engineer at a regional ISP, we used the open source cross-platform **Multi Router Traffic Grapher (MRTG)**, (https://en.wikipedia.org/wiki/Multi_Router_Traffic_Grapher) tool to check the traffic load on network links. We relied on the tool almost exclusively for traffic monitoring. I was really amazed at how good and useful an open source project could be. It was one of the first open source high-level network monitoring systems that abstracted the details of SNMP, the database, and HTML for network engineers. Then came the **Round-Robin Database Tool (RRDtool)**, (https://en.wikipedia.org/wiki/RRDtool). In its first release in 1999, it was referred to as "MRTG Done Right". It had greatly improved the database and poller performance in the backend.

Released in 2001, Cacti (https://en.wikipedia.org/wiki/Cacti_(software)) is an open source web-based network monitoring and graphing tool designed as an improved frontend for RRDtool. Because of the heritage of MRTG and RRDtool, you will notice a familiar graph layout, templates, and SNMP poller. As a packaged tool, the installation and usage will need to stay within the boundary of the tool itself. However, Cacti offers the custom data query feature that we can use Python for. In this section, we will see how we can use Python as an input method for Cacti.

Installation

Installation on Ubuntu is straightforward when using APT on the Ubuntu management VM:

```
$ sudo apt-get install cacti
```

It will trigger a series of installation and setup steps, including the MySQL database, web server (Apache or lighttpd), and various configuration tasks. Once installed, navigate to `http://<ip>/cacti` to get started. The last step is to log in with the default username and password (`admin/admin`); you will be prompted to change the password.

Once you are logged in, you can follow the documentation to add a device and associate it with a template. There is a Cisco router premade template that you can go with. Cacti has good documentation on `http://docs.cacti.net/` for adding a device and creating your first graph, so we will quickly look at some screenshots that you can expect to see:

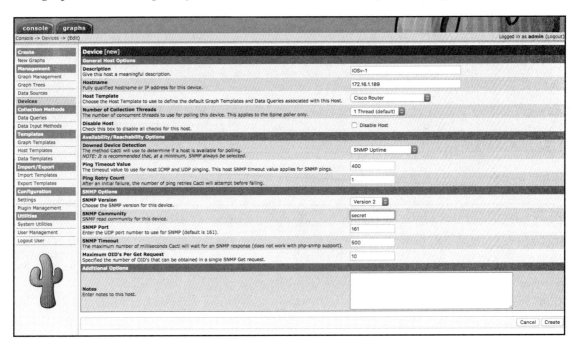

Network Monitoring with Python – Part 1

A sign indicating the SNMP communication is working is when you can see the device uptime:

You can add graphs to the device for interface traffic and other statistics:

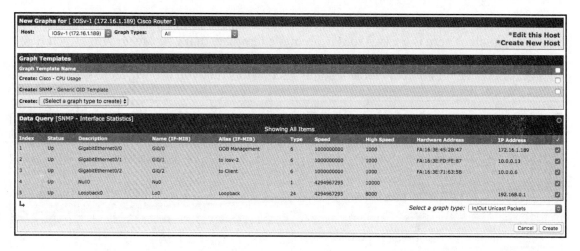

After some time, you will start seeing traffic, as shown here:

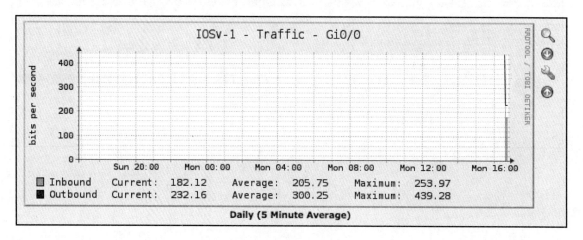

We are now ready to look at how to use Python scripts to extend Cacti's data gathering functionality.

Python script as an input source

There are two documents that we should read before we try to use our Python script as an input source:

- **Data input methods:** http://www.cacti.net/downloads/docs/html/data_input_methods.html
- **Making your scripts work with Cacti:** http://www.cacti.net/downloads/docs/html/making_scripts_work_with_cacti.html

One might wonder what the use cases for using Python script are as an extension for data inputs. One of the use cases would be to provide monitoring to resources that do not have a corresponding OID, for example, if we would like to know how to graph how many times the access list `permit_snmp` has allowed the host `172.16.1.173` for conducting an SNMP query. We know we can see the number of matches via the CLI:

```
iosv-1#sh ip access-lists permit_snmp | i 172.16.1.173
    10 permit 172.16.1.173 log (6362 matches)
```

However, chances are that there are no OIDs associated with this value (or we can just pretend that there are none). This is where we can use an external script to produce an output that can be consumed by the Cacti host.

We can reuse the Pexpect script we discussed in Chapter 2, *Low-Level Network Device Interactions*, `chapter1_1.py`. We will rename it to `cacti_1.py`. Everything should be familiar to the original script, except that we will execute the CLI command and save the output:

```
for device in devices.keys():
...
    child.sendline('sh ip access-lists permit_snmp | i 172.16.1.173')
    child.expect(device_prompt)
    output = child.before
...
```

The output in its raw form will appear as follows:

```
b'sh ip access-lists permit_snmp | i 172.16.1.173rn 10 permit 172.16.1.173 log (6428 matches)rn'
```

We will use the split() function for the string to only leave the number of matches and print them out on standard output in the script:

```
print(str(output).split('(')[1].split()[0])
```

To test this, we can see the number of increments by executing the script a number of times:

```
$ ./cacti_1.py
6428
$ ./cacti_1.py
6560
$ ./cacti_1.py
6758
```

We can make the script executable and put it into the default Cacti script location:

```
$ chmod a+x cacti_1.py
$ sudo cp cacti_1.py /usr/share/cacti/site/scripts/
```

The Cacti documentation, available at http://www.cacti.net/downloads/docs/html/how_to.html, provides detailed steps on how to add the script result to the output graph. These steps include adding the script as a data input method, adding the input method to a data source, and then creating a graph to be viewed:

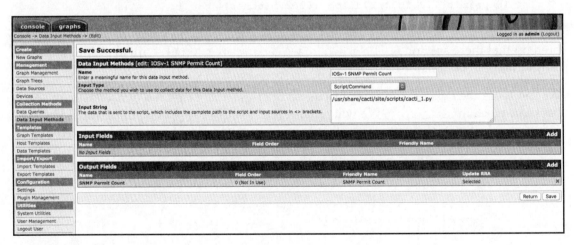

SNMP is a common way to provide network monitoring services to the devices. RRDtool with Cacti as the frontend provides a good platform to be used for all the network devices via SNMP.

Summary

In this chapter, we explored ways to perform network monitoring via SNMP. We configured SNMP-related commands on network devices and used our network management VM with SNMP poller to query the devices. We used the PySNMP module to simplify and automate our SNMP queries. We also learned how to save the query results in a flat file or database to be used for future examples.

Later in this chapter, we used two different Python visualization packages, namely Matplotlib and Pygal, to graph SNMP results. Each package has its distinct advantages. Matplotlib is a mature, feature-rich library that is widely used in data science projects. Pygal can natively generate SVG format graphs that are flexible and web-friendly. We saw how we can generate line and pie graphs that are relevant for network monitoring.

Toward the end of this chapter, we looked at an all-inclusive network monitoring tool named Cacti. It uses primarily SNMP for network monitoring, but we saw how we can use Python scripts as an input source to extend the platform's monitoring capabilities when SNMP OID is not available on the remote host.

In `Chapter 8`, *Network Monitoring with Python – Part 2*, we will continue to discuss the tools we can use to monitor our networks and gain insight into whether the network is behaving as expected. We will look at flow-based monitoring using NetFlow, sFlow, and IPFIX. We will also use tools such as Graphviz to visualize our network topology and detect any topological changes. Finally, we will use Elasticsearch, Logstash, and Kibana, commonly referred to as the ELK stack, to monitor network log data as well as other network-related input.

8
Network Monitoring with Python – Part 2

In Chapter 7, *Network Monitoring with Python – Part 1*, we used SNMP to query information from network devices. We did this by using an SNMP manager to query the SNMP agent residing on the network device. The SNMP information is structured in a hierarchy format with a specific object ID as the way to represent the value of the object. Most of the time, the value we care about is a number, such as CPU load, memory usage, or interface traffic. It's something we can graph against time to give us a sense of how the value has changed over time.

We can typically classify the SNMP approach as a `pull` method as we are constantly asking the device for a particular answer. This particular method adds burden to the device because it needs to spend a CPU cycle on the control plane to find answers from the subsystem, package the answer in an SNMP packet, and transport the answer back to the poller. If you have ever been to a family reunion where you have that one family member who keeps asking you the same questions over and over again, that would be analogous to the SNMP manager polling the managed node.

Over time, if we have multiple SNMP pollers querying the same device every 30 seconds (you would be surprised how often this happens), the management overhead would become substantial. In the same family reunion example we have given, instead of one family member, imagine there are many other people interrupting you every 30 seconds to ask you a question. I don't know about you, but I know I would be very annoyed even if it was a simple question (or worse if all of them are asking the same question).

Another way we can provide more efficient network monitoring is to reverse the relationship between the management station from a pull to a push model. In other words, the information can be pushed from the device toward the management station in an agreed-upon format. This concept is what flow-based monitoring is based on. In a flow-based model, the network device streams the traffic information, called flow, to the management station. The format can be the Cisco proprietary NetFlow (version 5 or version 9), the industry standard IPFIX, or the open source sFlow format. In this chapter, we will spend some time looking into NetFlow, IPFIX, and sFlow with Python.

Not all monitoring comes in the form of time series data. You can represent information such as network topology and Syslog in a time series format if you really want to, but, this is not ideal. We can use Python to check network topology information and see if the topology has changed over time. We can use tools, such as Graphviz, with a Python wrapper, to illustrate the topology. As already seen in `Chapter 6`, *Network Security with Python*, Syslog contains security information. In this chapter, we will look at using the ELK stack (Elasticsearch, Logstash, Kibana) as an efficient way to collect and index network log information.

Specifically, in this chapter, we will cover the following topics:

- Graphviz, which is an open source graph visualization software that can help us quickly and efficiently graph our network
- Flow-based monitoring, such as NetFlow, IPFIX, and sFlow
- Using ntop to visualize the flow information
- Using Elasticsearch to index and analyze our collected data

Let's start by looking at how to use Graphviz as a tool to monitor network topology changes.

Graphviz

Graphviz is an open source graph visualization software. Imagine if we have to describe our network topology to a colleague without the benefit of a picture. We might say, our network consists of three layers: core, distribution, and access. The core layer comprises two routers for redundancy, and both of the routers are full-meshed toward the four distribution routers; the distribution routers are also full-meshed toward the access routers. The internal routing protocol is OSPF, and externally, we use BGP for peering with our service provider. While this description lacks some details, it is probably enough for your colleague to paint a pretty good high-level picture of your network.

Graphviz works similarly to the process by describing the graph in the text format that Graphviz can understand, then we can feed the file to the Graphviz program to construct the graph for us. Here, the graph is described in a text format called DOT (https://en.wikipedia.org/wiki/DOT_(graph_description_language)) and Graphviz renders the graph based on the description. Of course, because the computer lacks human imagination, the language has to be very precise and detailed.

 For Graphviz-specific DOT grammar definitions, take a look at http://www.graphviz.org/doc/info/lang.html.

In this section, we will use the **Link Layer Discovery Protocol** (**LLDP**) to query the device neighbors and create a network topology graph via Graphviz. Upon completing this extensive example, we will see how we can take something new, such as Graphviz, and combine it with things we have already learned to solve interesting problems.

Let's start by constructing the lab we will be using.

Lab setup

We will use VIRL to construct our lab. As in the previous chapters, we will put together a lab with multiple routers, a server, and a client. We will use five IOSv network nodes along with two server hosts:

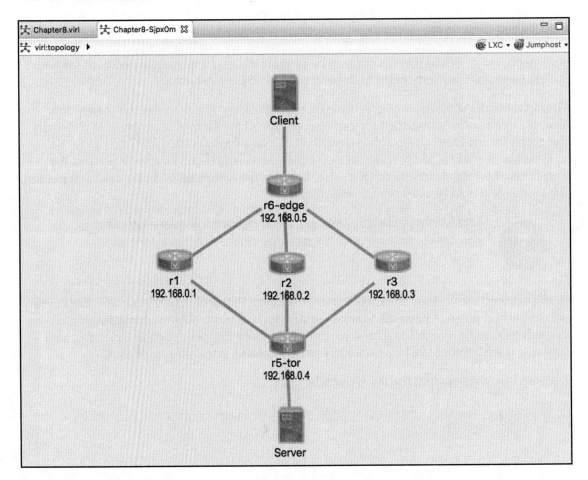

If you are wondering about our choice of IOSv as opposed to NX-OS or IOS-XR and the number of devices, here are a few points for you to consider when you build your own lab:

- Nodes virtualized by NX-OS and IOS-XR are much more memory-intensive than IOS
- The VIRL virtual manager I am using has 8 GB of RAM, which seems enough to sustain nine nodes but could be a bit unstable (nodes changing from reachable to unreachable at random)

- If you wish to use NX-OS, consider using NX-API or other API calls that would return structured data

For our example, we are going to use LLDP as the protocol for link layer neighbor discovery because it is vendor-neutral. Note that VIRL provides an option to automatically enable CDP, which can save you some time and is similar to LLDP in functionality; however, it is a Cisco proprietary technology so we will disable it for our lab:

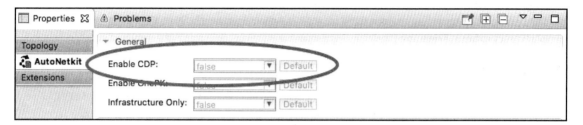

Once the lab is up and running, proceed to installing the necessary software packages.

Installation

Graphviz can be obtained via `apt`:

```
$ sudo apt-get -y install graphviz
```

After the installation is complete, note that verification is performed by using the `dot` command:

```
$ dot -V
dot - graphviz version 2.38.0 (20140413.2041)~
```

We will use the Python wrapper for Graphviz, so let's install it now while we are at it:

```
$ sudo pip install graphviz #Python 2
$ sudo pip3 install graphviz

$ python3
Python 3.5.2 (default, Nov 23 2017, 16:37:01)
[GCC 5.4.0 20160609] on linux
Type "help", "copyright", "credits" or "license" for more information.
>>> import graphviz
>>> graphviz.__version__
'0.8.4'
>>> exit()
```

Let's take a look at how we can use the software.

Graphviz examples

Like most popular open source projects, the documentation of Graphviz (http://www.graphviz.org/Documentation.php) is extensive. The challenge for someone new to the software is often where to start. For our purpose, we will focus on the dot graph, which draws directed graphs as hierarchies (not to be confused with the DOT language, which is a graph description language).

Let's start with some of the basic concepts:

- Nodes represent our network entities, such as routers, switches, and servers
- The edge represents the link between the network entities
- The graph, nodes, and edges each have attributes (https://www.graphviz.org/doc/info/attrs.html) that can be tweaked
- After describing the network, we can output the network graph (https://www.graphviz.org/doc/info/output.html) in either PNG, JPEG, or PDF format

Our first example is an undirected dot graph consisting of four nodes (core, distribution, access1, and access2). The edges, represented by the dash - sign, join the core node to the distribution node, as well as the distribution node to both of the access nodes:

```
$ cat chapter8_gv_1.gv
graph my_network {
 core -- distribution;
 distribution -- access1;
 distribution -- access2;
}
```

The graph can be output in the dot -T<format> source -o <output file> command line:

```
$ dot -Tpng chapter8_gv_1.gv -o output/chapter8_gv_1.png
```

The resultant graph can be viewed from the following output folder:

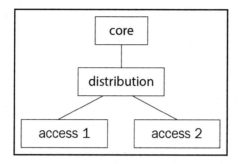

 Just like Chapter 7, *Network Monitoring with Python – Part 1*, it might be easier to work in the Linux desktop window while working with these graphs so you can see the graphs right away.

Note that we can use a directional graph by specifying the graph as a digraph as well as using the arrow (->) sign to represent the edges. There are several attributes we can modify in the case of nodes and edges, such as the node shape, edge labels, and so on. The same graph can be modified as follows:

```
$ cat chapter8_gv_2.gv
digraph my_network {
  node [shape=box];
  size = "50 30";
  core -> distribution [label="2x10G"];
  distribution -> access1 [label="1G"];
  distribution -> access2 [label="1G"];
}
```

We will output the file in PDF this time:

```
$ dot -Tpdf chapter8_gv_2.gv -o output/chapter8_gv_2.pdf
```

Take a look at the directional arrows in the new graph:

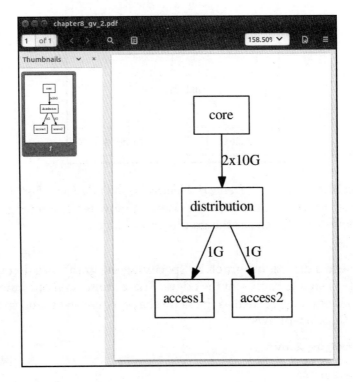

Now let's take a look at the Python wrapper around Graphviz.

Python with Graphviz examples

We can reproduce the same topology graph as before using the Python Graphviz package which we have installed:

```
$ python3
Python 3.5.2 (default, Nov 17 2016, 17:05:23)
>>> from graphviz import Digraph
>>> my_graph = Digraph(comment="My Network")
>>> my_graph.node("core")
>>> my_graph.node("distribution")
>>> my_graph.node("access1")
>>> my_graph.node("access2")
>>> my_graph.edge("core", "distribution")
>>> my_graph.edge("distribution", "access1")
>>> my_graph.edge("distribution", "access2")
```

The code basically produces what you would normally write in the DOT language but in a more Pythonic way. You can view the source of the graph before the graph generation:

```
>>> print(my_graph.source)
// My Network
digraph {
        core
        distribution
        access1
        access2
                core -> distribution
                distribution -> access1
                distribution -> access2
}
```

The graph can be rendered by the `render()` method; by default, the output format is PDF:

```
>>> my_graph.render("output/chapter8_gv_3.gv")
'output/chapter8_gv_3.gv.pdf'
```

The Python package wrapper closely mimics all the API options of Graphviz. You can find documentation about the options on the Graphviz Read the Docs website (http://graphviz.readthedocs.io/en/latest/index.html). You can also refer to the source code on GitHub for more information (https://github.com/xflr6/graphviz). We are now ready to use the tool to map out our network.

LLDP neighbor graphing

In this section, we will use the example of mapping out LLDP neighbors to illustrate a problem-solving pattern that has helped me over the years:

1. Modularize each task into smaller pieces, if possible. In our example, we can combine a few steps, but if we break them into smaller pieces, we will be able to reuse and improve them more easily.
2. Use an automation tool to interact with the network devices, but keep the more complex logic aside at the management station. For example, the router has provided an LLDP neighbor output that is a bit messy. In this case, we will stick with the working command and the output and use a Python script at the management station to parse out the output we need.

3. When presented with choices for the same task, pick the one that can be reused. In our example, we can use low-level Pexpect, Paramiko, or Ansible playbooks to query the routers. In my opinion, Ansible is a more reusable option, so that is what I have picked.

To get started, since LLDP is not enabled on the routers by default, we will need to configure them on the devices first. By now, we know we have a number of options to choose from; in this case, I chose the Ansible playbook with the `ios_config` module for the task. The `hosts` file consists of five routers:

```
$ cat hosts
[devices]
r1 ansible_hostname=172.16.1.218
r2 ansible_hostname=172.16.1.219
r3 ansible_hostname=172.16.1.220
r5-tor ansible_hostname=172.16.1.221
r6-edge ansible_hostname=172.16.1.222
```

The `cisco_config_lldp.yml` playbook consists of one play with variables embedded in the playbook to configure LLDP:

```
<skip>
 vars:
   cli:
     host: "{{ ansible_hostname }}"
     username: cisco
     password: cisco
     transport: cli tasks:
  - name: enable LLDP run
      ios_config:
        lines: lldp run
        provider: "{{ cli }}"
<skip>
```

After a few seconds, to allow LLDP exchange, we can verify that LLDP is indeed active on the routers:

```
$ ansible-playbook -i hosts cisco_config_lldp.yml

PLAY [Enable LLDP]
************************************************************
...
PLAY RECAP
****************************************************************
r1   : ok=2  changed=1  unreachable=0  failed=0
r2   : ok=2  changed=1  unreachable=0  failed=0
r3   : ok=2  changed=1  unreachable=0  failed=0
```

Chapter 8

```
r5-tor    : ok=2    changed=1    unreachable=0    failed=0
r6-edge   : ok=2    changed=1    unreachable=0    failed=0

## SSH to R1 for verification
r1#show lldp neighbors

Capability codes: (R) Router, (B) Bridge, (T) Telephone, (C) DOCSIS Cable
Device (W) WLAN Access Point, (P) Repeater, (S) Station, (O) Other

Device ID Local Intf Hold-time Capability Port ID
r2.virl.info Gi0/0 120 R Gi0/0
r3.virl.info Gi0/0 120 R Gi0/0
r5-tor.virl.info Gi0/0 120 R Gi0/0
r5-tor.virl.info Gi0/1 120 R Gi0/1
r6-edge.virl.info Gi0/2 120 R Gi0/1
r6-edge.virl.info Gi0/0 120 R Gi0/0

Total entries displayed: 6
```

In the output, you will see that `G0/0` is configured as the MGMT interface; therefore, you will see LLDP peers as if they are on a flat management network. What we really care about is the `G0/1` and `G0/2` interfaces connected to other peers. This knowledge will come in handy as we prepare to parse the output and construct our topology graph.

Information retrieval

We can now use another Ansible playbook, namely `cisco_discover_lldp.yml`, to execute the LLDP command on the device and copy the output of each device to a `tmp` directory:

```
<skip>
 tasks:
   - name: Query for LLDP Neighbors
     ios_command:
       commands: show lldp neighbors
       provider: "{{ cli }}"
<skip>
```

The `./tmp` directory now consists of all the routers' output (showing LLDP neighbors) in its own file:

```
$ ls -l tmp/
total 20
-rw-rw-r-- 1 echou echou 630 Mar 13 17:12 r1_lldp_output.txt
-rw-rw-r-- 1 echou echou 630 Mar 13 17:12 r2_lldp_output.txt
-rw-rw-r-- 1 echou echou 701 Mar 12 12:28 r3_lldp_output.txt
```

```
-rw-rw-r-- 1 echou echou 772 Mar 12 12:28 r5-tor_lldp_output.txt
-rw-rw-r-- 1 echou echou 630 Mar 13 17:12 r6-edge_lldp_output.txt
```

The `r1_lldp_output.txt` content is the `output.stdout_lines` variable from our Ansible playbook:

```
$ cat tmp/r1_lldp_output.txt
[["Capability codes:", " (R) Router, (B) Bridge, (T) Telephone, (C) DOCSIS
Cable Device", " (W) WLAN Access Point, (P) Repeater, (S) Station, (O)
Other", "", "Device ID Local Intf Hold-time Capability Port ID",
"r2.virl.info Gi0/0 120 R Gi0/0", "r3.virl.info Gi0/0 120 R Gi0/0", "r5-
tor.virl.info Gi0/0 120 R Gi0/0", "r5-tor.virl.info Gi0/1 120 R Gi0/1",
"r6-edge.virl.info Gi0/0 120 R Gi0/0", "", "Total entries displayed: 5",
""]]
```

Python parser script

We can now use a Python script to parse the LLDP neighbor output from each device and construct a network topology graph from the results. The purpose is to automatically check the device to see whether any of the LLDP neighbors have disappeared due to link failure or other issues. Let's take a look at the `cisco_graph_lldp.py` file and see how that is done.

We start with the necessary imports of the packages: an empty list that we will populate with tuples of node relationships. We also know that `Gi0/0` on the devices are connected to the management network; therefore, we are only searching for `Gi0/[1234]` as our regular expression pattern in the `show LLDP neighbors` output:

```
import glob, re
from graphviz import Digraph, Source
pattern = re.compile('Gi0/[1234]')
device_lldp_neighbors = []
```

We will use the `glob.glob()` method to traverse the `./tmp` directory of all the files, parse out the device name, and find the neighbors that the device is connected to. There are some embedded print statements in the script that we can comment out for the final version; if the statements were uncommented, we can see the parsed result:

```
device: r1
 neighbors: r5-tor
 neighbors: r6-edge
device: r5-tor
 neighbors: r2
 neighbors: r3
```

```
  neighbors: r1
device: r2
  neighbors: r5-tor
  neighbors: r6-edge
device: r3
  neighbors: r5-tor
  neighbors: r6-edge
device: r6-edge
  neighbors: r2
  neighbors: r3
  neighbors: r1
```

The fully populated edge list contains tuples that consist of the device and its neighbors:

```
Edges: [('r1', 'r5-tor'), ('r1', 'r6-edge'), ('r5-tor', 'r2'), ('r5-tor',
'r3'), ('r5-tor', 'r1'), ('r2', 'r5-tor'), ('r2', 'r6-edge'), ('r3', 'r5-
tor'), ('r3', 'r6-edge'), ('r6-edge', 'r2'), ('r6-edge', 'r3'), ('r6-edge',
'r1')]
```

We can now construct the network topology graph using the Graphviz package. The most important part is the unpacking of the tuples that represent the edge relationship:

```
my_graph = Digraph("My_Network")
<skip>
# construct the edge relationships
for neighbors in device_lldp_neighbors:
    node1, node2 = neighbors
    my_graph.edge(node1, node2)
```

If we were to print out the resulting source dot file, it would be an accurate representation of our network:

```
digraph My_Network {
    r1 -> "r5-tor"
    r1 -> "r6-edge"
    "r5-tor" -> r2
    "r5-tor" -> r3
    "r5-tor" -> r1
    r2 -> "r5-tor"
    r2 -> "r6-edge"
    r3 -> "r5-tor"
    r3 -> "r6-edge"
    "r6-edge" -> r2
    "r6-edge" -> r3
    "r6-edge" -> r1
}
```

Sometimes, it is confusing to see the same link twice; for example, the r2 to r5-tor link appeared twice in the previous diagram for each of the directions of the link. As network engineers, we understand that sometimes a fault in the physical link will result in a unidirectional link, which we want to see.

If we were to graph the diagram as is, the placement of the nodes would be a bit funky. The placement of the nodes is auto-rendered. The following diagram illustrates the rendering in a default layout as well as the neato layout, namely a digraph (My_Network, engine='neato'):

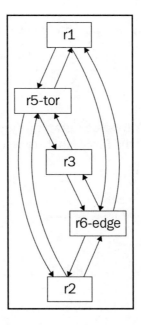

The neato layout represents an attempt to draw undirected graphs with even less hierarchy:

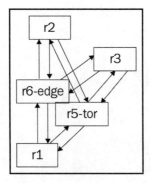

Chapter 8

Sometimes, the default layout presented by the tool is just fine, especially if your goal is to detect faults as opposed to making it visually appealing. However, in this case, let's see how we can insert raw DOT language knobs into the source file. From research, we know that we can use the `rank` command to specify the level where some nodes can stay on the same level. However, there is no option presented in the Graphviz Python API. Luckily, the dot source file is just a string, which we can insert as raw dot comments using the `replace()` method with the following:

```
source = my_graph.source
original_text = "digraph My_Network {"
new_text = 'digraph My_Network {n{rank=same Client "r6-edge"}n{rank=same r1 r2 r3}n'
new_source = source.replace(original_text, new_text)
new_graph = Source(new_source)new_graph.render("output/chapter8_lldp_graph.gv")
```

The end result is a new source that we can render the final topology graph from:

```
digraph My_Network {
{rank=same Client "r6-edge"}
{rank=same r1 r2 r3}
                Client -> "r6-edge"
                "r5-tor" -> Server
                r1 -> "r5-tor"
                r1 -> "r6-edge"
                "r5-tor" -> r2
                "r5-tor" -> r3
                "r5-tor" -> r1
                r2 -> "r5-tor"
                r2 -> "r6-edge"
                r3 -> "r5-tor"
                r3 -> "r6-edge"
                "r6-edge" -> r2
                "r6-edge" -> r3
                "r6-edge" -> r1
}
```

The graph is now good to go:

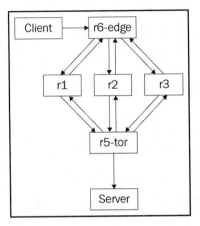

Final playbook

We are now ready to incorporate this new parser script back into our playbook. We can now add the additional task of rendering the output with graph generation in `cisco_discover_lldp.yml`:

```
tasks:
  - name: Query for LLDP Neighbors
    ios_command:
      commands: show lldp neighbors
      provider: "{{ cli }}"

    register: output

  - name: show output
    debug:
      var: output

  - name: copy output to file
    copy: content="{{ output.stdout_lines }}" dest="./tmp/{{ inventory_hostname }}_lldp_output.txt"

  - name: Execute Python script to render output
    command: ./cisco_graph_lldp.py
```

This playbook will now include four tasks, covering the end-to-end process of executing the `show lldp` command on the Cisco devices, displaying the output on the screen, copying the output to a separate file, and then rendering the output via a Python script.

Chapter 8

The playbook can now be scheduled to run regularly via cron or other means. It will automatically query the devices for LLDP neighbors and construct the graph, and the graph will represent the current topology as known by the routers.

We can test this by shutting down the Gi0/1 and Go0/2 interfaces on r6-edge. When the LLDP neighbor passes the hold timer, they will disappear from the LLDP table on r6-edge:

```
r6-edge#sh lldp neighbors
...
Device ID Local Intf Hold-time Capability Port ID
r2.virl.info Gi0/0 120 R Gi0/0
r3.virl.info Gi0/3 120 R Gi0/2
r3.virl.info Gi0/0 120 R Gi0/0
r5-tor.virl.info Gi0/0 120 R Gi0/0
r1.virl.info Gi0/0 120 R Gi0/0

Total entries displayed: 5
```

If we execute the playbook, the graph will automatically show that r6-edge only connects to r3 and we can start to troubleshoot why that is the case:

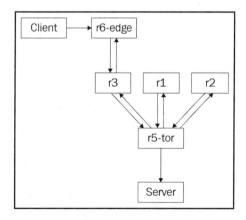

This is a relatively long example. We used the tools we have learned so far in the book—Ansible and Python—to modularize and break tasks into reusable pieces. We then used a new tool, namely Graphviz, to help monitor the network for non-time series data, such as network topology relationships.

[259]

Flow-based monitoring

As mentioned in the chapter introduction, besides polling technology, such as SNMP, we can also use a push strategy, which allows the device to push network information toward the management station. NetFlow and its closely associated cousins, IPFIX and sFlow, are examples of such information pushed from the direction of the network device toward the management station. We can make the argument that the `push` method is more sustainable since the network device is inherently in charge of allocating the necessary resources to push the information. If the device CPU is busy, for example, it can choose to skip the flow export process in favor of routing packets, which is what we want.

A flow, as defined by IETF (`https://www.ietf.org/proceedings/39/slides/int/ip1394-background/tsld004.htm`), is a sequence of packets moving from an application sending something to the application receiving it. If we refer back to the OSI model, a flow is what constitutes a single unit of communication between two applications. Each flow comprises a number of packets; some flows have more packets (such as a video stream), while some have just a few (such as an HTTP request). If you think about flows for a minute, you'll notice that routers and switches might care about packets and frames, but the application and user usually care more about the network flows.

Flow-based monitoring usually refers to NetFlow, IPFIX, and sFlow:

- **NetFlow**: NetFlow v5 is a technology where the network device caches flow entries and aggregate packets by matching the set of tuples (source interface, source IP/port, destination IP/port, and so on). Here, once a flow is completed, the network device exports the flow characteristics, including total bytes and packet counts in the flow, to the management station.
- **IPFIX**: IPFIX is the proposed standard for structured streaming and is similar to NetFlow v9, also known as Flexible NetFlow. Essentially, it is a definable flow export, which allows the user to export nearly anything that the network device knows about. The flexibility often comes at the expense of simplicity compared to NetFlow v5. The configuration of IPFIX is more complex than the traditional NetFlow v5. Additional complexity makes it less ideal for introductory learning. However, once you are familiar with NetFlow v5, you will be able to parse IPFIX as long as you match the template definition.

- **sFlow**: sFlow actually has no notion of a flow or packet aggregation by itself. It performs two types of sampling of packets. It randomly samples one out of *n* packets/applications and has a time-based sampling counter. It sends the information to the management station, and the station derives the network flow information by referring to the type of packet sample received along with the counters. As it doesn't perform any aggregation on the network device, you can argue that sFlow is more scalable than NetFlow and IPFIX.

The best way to learn about each one of these is probably to dive right into examples.

NetFlow parsing with Python

We can use Python to parse the NetFlow datagram being transported on the wire. This gives us a way to look at the NetFlow packet in detail as well as troubleshoot any NetFlow issues when it is not working as expected.

First, let's generate some traffic between the client and server across the VIRL network. We can use the built-in HTTP server module from Python to quickly launch a simple HTTP server on the VIRL host acting as the server:

```
cisco@Server:~$ python3 -m http.server
Serving HTTP on 0.0.0.0 port 8000 ...
```

> For Python 2, the module is named `SimpleHTTPServer`; for example, `python2 -m SimpleHTTPServer`.

We can create a short `while` loop in a Python script to continuously send `HTTP GET` to the web server on the client:

```
sudo apt-get install python-pip python3-pip
sudo pip install requests
sudo pip3 install requests

$ cat http_get.py
import requests, time
while True:
    r = requests.get('http://10.0.0.5:8000')
    print(r.text)
    time.sleep(5)
```

The client should get a very plain HTML page:

```
cisco@Client:~$ python3 http_get.py
<!DOCTYPE html PUBLIC "-//W3C//DTD HTML 3.2 Final//EN"><html>
<title>Directory listing for /</title>
<body>
...
</body>
</html>
```

We should also see the requests continuously coming in from the client every five seconds:

```
cisco@Server:~$ python3 -m http.server
Serving HTTP on 0.0.0.0 port 8000 ...
10.0.0.9 - - [15/Mar/2017 08:28:29] "GET / HTTP/1.1" 200 -
10.0.0.9 - - [15/Mar/2017 08:28:34] "GET / HTTP/1.1" 200 -
```

We can export NetFlow from any of the devices, but since `r6-edge` is the first hop for the client host, we will have this router export NetFlow to the management host at port `9995`.

In this example, we use only one device for demonstration; therefore, we manually configure it with the necessary commands. In the next section, when we enable NetFlow on all the devices, we will use an Ansible playbook to configure all the routers at once.

The following configurations are necessary for exporting NetFlow on the Cisco IOS devices:

```
!
ip flow-export version 5
ip flow-export destination 172.16.1.173 9995 vrf Mgmt-intf
!
interface GigabitEthernet0/4
 description to Client
 ip address 10.0.0.10 255.255.255.252
 ip flow ingress
 ip flow egress
...
!
```

Next, let's take a look at the Python parser script.

Python socket and struct

The script, netFlow_v5_parser.py, was modified from Brian Rak's blog post at http://blog.devicenull.org/2013/09/04/python-netflow-v5-parser.html. The modification was mainly for Python 3 compatibility as well as parsing additional NetFlow version 5 fields. The reason we choose NetFlow v5 instead of NetFlow v9 is that v9 is more complex and uses templates to map out the fields, making it more difficult to learn in an introductory session. However, since NetFlow version 9 is an extended format of the original NetFlow version 5, all the concepts we introduced in this section are applicable to it.

Because NetFlow packets are represented in bytes over the wire, we will use the Python struct module included in the standard library to convert bytes into native Python data types.

> You'll find more information about the two modules at https://docs.python.org/3.5/library/socket.html and https://docs.python.org/3.5/library/struct.html.

We will start by using the socket module to bind and listen for the UDP datagrams. With socket.AF_INET, we intend on listing for the IPv4 address sockets; with socket.SOCK_DGRAM, we specify that we'll see the UDP datagram:

```
sock = socket.socket(socket.AF_INET, socket.SOCK_DGRAM)
sock.bind(('0.0.0.0', 9995))
```

We will start a loop and retrieve information off the wire 1,500 bytes at a time:

```
while True:
        buf, addr = sock.recvfrom(1500)
```

The following line is where we begin to deconstruct or unpack the packet. The first argument of !HH specifies the network's big-endian byte order with the exclamation sign (big-endian) as well as the format of the C type (H = 2 byte unsigned short integer):

```
(version, count) = struct.unpack('!HH',buf[0:4])
```

The first four bytes include the version and the number of flows exported in this packet. If you do not remember the NetFlow version 5 header off the top of your head (that was a joke, by the way; I only read the header when I want to fall asleep quickly), here is a quick glance:

Table B-3 Version 5 Header Format

Bytes	Contents	Description
0-1	version	NetFlow export format version number
2-3	count	Number of flows exported in this packet (1-30)
4-7	SysUptime	Current time in milliseconds since the export device booted
8-11	unix_secs	Current count of seconds since 0000 UTC 1970
12-15	unix_nsecs	Residual nanoseconds since 0000 UTC 1970
16-19	flow_sequence	Sequence counter of total flows seen
20	engine_type	Type of flow-switching engine
21	engine_id	Slot number of the flow-switching engine
22-23	sampling_interval	First two bits hold the sampling mode; remaining 14 bits hold value of sampling interval

NetFlow v5 header (source: http://www.cisco.com/c/en/us/td/docs/net_mgmt/netflow_collection_engine/3-6/user/guide/format.html#wp1006108)

The rest of the header can be parsed accordingly, depending on the byte location and data type:

```
(sys_uptime, unix_secs, unix_nsecs, flow_sequence) = struct.unpack('!IIII', buf[4:20])
(engine_type, engine_id, sampling_interval) = struct.unpack('!BBH', buf[20:24])
```

The `while` loop that follows will fill the `nfdata` dictionary with the flow record that unpacks the source address and port, destination address and port, packet count, and byte count, and print the information out on the screen:

```
for i in range(0, count):
    try:
        base = SIZE_OF_HEADER+(i*SIZE_OF_RECORD)
        data = struct.unpack('!IIIIHH',buf[base+16:base+36])
        input_int, output_int = struct.unpack('!HH', buf[base+12:base+16])
        nfdata[i] = {}
        nfdata[i]['saddr'] = inet_ntoa(buf[base+0:base+4])
        nfdata[i]['daddr'] = inet_ntoa(buf[base+4:base+8])
        nfdata[i]['pcount'] = data[0]
        nfdata[i]['bcount'] = data[1]
...
```

The output of the script allows you to visualize the header as well as the flow content at a glance:

```
Headers:
NetFlow Version: 5
Flow Count: 9
System Uptime: 290826756
Epoch Time in seconds: 1489636168
Epoch Time in nanoseconds: 401224368
Sequence counter of total flow: 77616
0 192.168.0.1:26828 -> 192.168.0.5:179 1 packts 40 bytes
1 10.0.0.9:52912 -> 10.0.0.5:8000 6 packts 487 bytes
2 10.0.0.9:52912 -> 10.0.0.5:8000 6 packts 487 bytes
3 10.0.0.5:8000 -> 10.0.0.9:52912 5 packts 973 bytes
4 10.0.0.5:8000 -> 10.0.0.9:52912 5 packts 973 bytes
5 10.0.0.9:52913 -> 10.0.0.5:8000 6 packts 487 bytes
6 10.0.0.9:52913 -> 10.0.0.5:8000 6 packts 487 bytes
7 10.0.0.5:8000 -> 10.0.0.9:52913 5 packts 973 bytes
8 10.0.0.5:8000 -> 10.0.0.9:52913 5 packts 973 bytes
```

Note that, in NetFlow version 5, the size of the record is fixed at 48 bytes; therefore, the loop and script are relatively straightforward. However, in the case of NetFlow version 9 or IPFIX, after the header, there is a template FlowSet (http://www.cisco.com/en/US/technologies/tk648/tk362/technologies_white_paper09186a00800a3db9.html) that specifies the field count, field type, and field length. This allows the collector to parse the data without knowing the data format in advance.

By parsing the NetFlow data in a script, we gained a solid understanding of the fields, but this is very tedious and hard to scale. As you may have guessed, there are other tools that save us the problem of parsing NetFlow records one by one. Let's look at one such tool, called **ntop**, in the coming section.

ntop traffic monitoring

Just like the PySNMP script in Chapter 7, *Network Monitoring with Python – Part 1*, and the NetFlow parser script in this chapter, we can use Python scripts to handle low-level tasks on the wire. However, there are tools such as Cacti, which is an all-in-one open source package, that include data collection (poller), data storage (RRD), and a web frontend for visualization. These tools can save you a lot of work by packing the frequently used features and software in one package.

In the case of NetFlow, there are a number of open source and commercial NetFlow collectors you can choose from. If you do a quick search for top N open source NetFlow analyzers, you will see a number of comparison studies for different tools. Each one of them has its own strong and weak points; which one to use is really a matter of preference, platform, and your appetite for customization. I would recommend choosing a tool that would support both v5 and v9, and potentially sFlow as well. A secondary consideration would be if the tool is written in a language that you can understand; I imagine having Python extensibility would be a nice thing.

Two of the open source NetFlow tools that I like and have used before are NfSen (with NFDUMP as the backend collector) and ntop (or ntopng). Between the two of them, ntop is the better-known traffic analyzer; it runs on both Windows and Linux platforms and integrates well with Python. Therefore, let's use ntop as an example in this section.

The installation of our Ubuntu host is straightforward:

```
$ sudo apt-get install ntop
```

The installation process will prompt for the necessary interface for listening and setting the administrator password. By default, the ntop web interface listens on port 3000, while the probe listens on UDP port 5556. On the network device, we need to specify the location of the NetFlow exporter:

```
!
ip flow-export version 5
ip flow-export destination 172.16.1.173 5556 vrf Mgmt-intf
!
```

Chapter 8

By default, IOSv creates a VRF called Mgmt-intf and places Gi0/0 under VRF.

We will also need to specify the direction of traffic exports, such as ingress or egress, under the interface configuration:

```
!
interface GigabitEthernet0/0
...
 ip flow ingress
 ip flow egress
...
```

For your reference, I have included the Ansible playbook, cisco_config_netflow.yml, to configure the lab device for the NetFlow export.

The r5-tor and r6-edge have two interfaces more than r1, r2, and r3 do.

Execute the playbook and make sure the changes were applied properly on the devices:

```
$ ansible-playbook -i hosts cisco_config_netflow.yml

TASK [configure netflow export station]
****************************************
changed: [r1]
changed: [r3]
changed: [r2]
changed: [r5-tor]
changed: [r6-edge]

TASK [configure flow export on Gi0/0]
****************************************
changed: [r2]
changed: [r1]
changed: [r6-edge]
changed: [r5-tor]
changed: [r3]
...
PLAY RECAP
****************************************************************
r1       : ok=4    changed=4    unreachable=0    failed=0
r2       : ok=4    changed=4    unreachable=0    failed=0
```

[267]

```
r3 : ok=4 changed=4 unreachable=0 failed=0
r5-tor : ok=6 changed=6 unreachable=0 failed=0
r6-edge : ok=6 changed=6 unreachable=0 failed=0

##Checking r2 for NetFlow configuration
r2#sh run | i flow
 ip flow ingress
 ip flow egress
 ip flow ingress
 ip flow egress
 ip flow ingress
 ip flow egress
ip flow-export version 5
ip flow-export destination 172.16.1.173 5556 vrf Mgmt-intf
```

Once everything is set up, you can check the ntop web interface for local IP traffic:

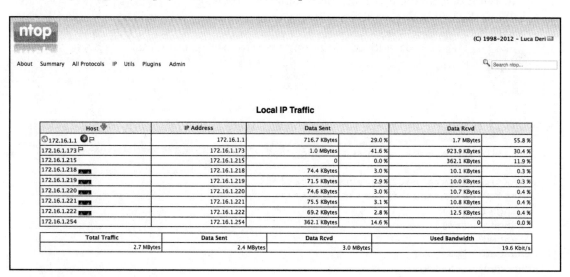

[268]

One of the most often used features of ntop is using it to look at the top talker graph:

The ntop reporting engine is written in C; it is fast and efficient, but the need to have adequate knowledge of C in order to do something as simple as change the web frontend does not fit the modern agile development mindset.

After a few false starts with Perl in the mid-2000s, the good folks at ntop finally settled on embedding Python as an extensible scripting engine. Let's take a look.

ns with Python – Part 2

Python extension for ntop

We can use Python to extend ntop through the ntop web server. The ntop web server can execute Python scripts. At a high level, the scripts will perform the following:

- Methods to access the state of ntop
- The Python CGI module to process forms and URL parameters
- Making templates that generate dynamic HTML pages
- Each Python script can read from `stdin` and print out `stdout/stderr`
- The `stdout` script is the returned HTTP page

There are several resources that come in handy with the Python integration. Under the web interface, you can click on **About** | **Show Configuration** to see the Python interpreter version as well as the directory for your Python script:

Run time/Internal	
Web server URL	http://any:3000
GDBM version	GDBM version 1.8.3. 10/15/2002 (built Nov 16 2014 23:11:58)
Embedded Python	2.7.12 (default, Nov 19 2016, 06:48:10) [GCC 5.4.0 20160609]

Python version

You can also check the various directories where the Python script should reside:

Directory (search) order	
Data Files	/usr/share/ntop /usr/local/share/ntop
Config Files	/usr/share/ntop /usr/local/etc/ntop /etc
Plugins	./plugins /usr/lib/ntop/plugins /usr/local/lib/ntop/plugins

Plugin directories

Under **About** | **Online Documentation** | **Python ntop Engine**, there are links for the Python API as well as the tutorial:

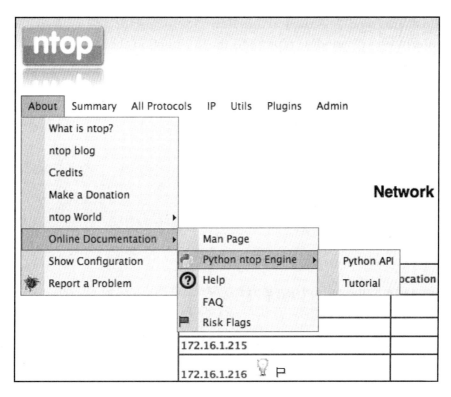

Python ntop documentation

As mentioned, the `ntop` web server directly executes the Python script placed under the designated directory:

```
$ pwd
/usr/share/ntop/python
```

We will place our first script, namely `chapter8_ntop_1.py`, in the directory. The Python CGI module processes forms and parses URL parameters:

```
# Import modules for CGI handling
import cgi, cgitb
import ntop

# Parse URL
cgitb.enable();
```

Network Monitoring with Python – Part 2

`ntop` implements three Python modules; each one of them has a specific purpose:

- **ntop**: This module interacts with the `ntop` engine
- **Host**: This module is used to drill down into a specific host's information
- **Interfaces**: This module represents the information about the localhost interfaces

In our script, we will use the `ntop` module to retrieve the `ntop` engine information as well as use the `sendString()` method to send the HTML body text:

```
form = cgi.FieldStorage();
name = form.getvalue('Name', default="Eric")

version = ntop.version()
os = ntop.os()
uptime = ntop.uptime()

ntop.printHTMLHeader('Mastering Python Networking', 1, 0)
ntop.sendString("Hello, "+ name +"<br>")
ntop.sendString("Ntop Information: %s %s %s" % (version, os, uptime))
ntop.printHTMLFooter()
```

We will execute the Python script using `http://<ip>:3000/python/<script name>`. Here is the result of our `chapter8_ntop_1.py` script:

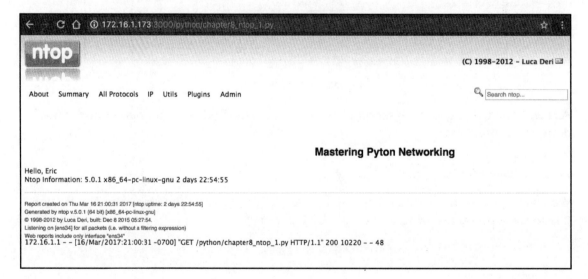

We can look at another example that interacts with the interface module, `chapter8_ntop_2.py`. We will use the API to iterate through the interfaces:

```
import ntop, interface, json

ifnames = []
try:
    for i in range(interface.numInterfaces()):
        ifnames.append(interface.name(i))

except Exception as inst:
    print type(inst) # the exception instance
    print inst.args # arguments stored in .args
    print inst # __str__ allows args to printed directly
...
```

The resulting page will display the ntop interfaces:

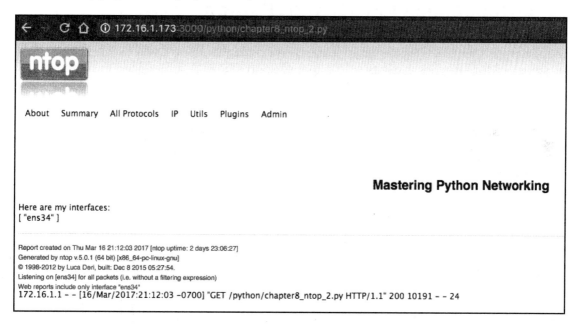

Besides the community version, ntop also offers a few commercial products that you can choose from. With the active open source community, commercial backing, and Python extensibility, ntop is a good choice for your NetFlow monitoring needs.

Next, let's take a look at NetFlow's cousin: sFlow.

sFlow

sFlow, which stands for sampled flow, was originally developed by InMon (http://www.inmon.com) and later standardized by way of RFC. The current version is v5. Many in the industry believe the primary advantage of sFlow is its scalability. sFlow uses random one in n packets flow samples along with the polling interval of counter samples to derive an estimate of the traffic; this is less CPU-intensive than NetFlow for the network devices. sFlow's statistical sampling is integrated with the hardware and provides real-time, raw exports.

For scalability and competitive reasons, sFlow is generally preferred over NetFlow for newer vendors, such as Arista Networks, Vyatta, and A10 Networks. While Cisco supports sFlow on its Nexus line of products, sFlow is generally *not* supported on Cisco platforms.

SFlowtool and sFlow-RT with Python

Unfortunately, at this point, sFlow is something that our VIRL lab devices do not support (not even with the NX-OSv virtual switches). You can either use a Cisco Nexus 3000 switch or other vendor switches, such as Arista, that support sFlow. Another good option for the lab is to use an Arista vEOS virtual instance. I happen to have access to a Cisco Nexus 3048 switch running 7.0 (3), which I will be using for this section as the sFlow exporter.

The configuration of Cisco Nexus 3000 for sFlow is straightforward:

```
Nexus-2# sh run | i sflow
feature sflow
sflow max-sampled-size 256
sflow counter-poll-interval 10
sflow collector-ip 192.168.199.185 vrf management
sflow agent-ip 192.168.199.148
sflow data-source interface Ethernet1/48
```

The easiest way to ingest sFlow is to use sflowtool. For installation instructions, refer to the document at http://blog.sflow.com/2011/12/sflowtool.html:

```
$ wget http://www.inmon.com/bin/sflowtool-3.22.tar.gz
$ tar -xvzf sflowtool-3.22.tar.gz
$ cd sflowtool-3.22/
$ ./configure
$ make
$ sudo make install
```

After the installation, you can launch `sflowtool` and look at the datagram Nexus 3048 is sending on the standard output:

```
$ sflowtool
startDatagram =================================
datagramSourceIP 192.168.199.148
datagramSize 88
unixSecondsUTC 1489727283
datagramVersion 5
agentSubId 100
agent 192.168.199.148
packetSequenceNo 5250248
sysUpTime 4017060520
samplesInPacket 1
startSample ----------------------
sampleType_tag 0:4
sampleType COUNTERSSAMPLE
sampleSequenceNo 2503508
sourceId 2:1
counterBlock_tag 0:1001
5s_cpu 0.00
1m_cpu 21.00
5m_cpu 20.80
total_memory_bytes 3997478912
free_memory_bytes 1083838464
endSample   ----------------------
endDatagram =================================
```

There are a number of good usage examples on the `sflowtool` GitHub repository (https://github.com/sflow/sflowtool); one of them is to use a script to receive the `sflowtool` input and parse the output. We can use a Python script for this purpose. In the `chapter8_sflowtool_1.py` example, we will use `sys.stdin.readline` to receive the input and use a regular expression search to print out only the lines containing the word `agent` when we see the sFlow packets:

```
import sys, re
for line in iter(sys.stdin.readline, ''):
    if re.search('agent ', line):
        print(line.strip())
```

The script can be piped to `sflowtool`:

```
$ sflowtool | python3 chapter8_sflowtool_1.py
agent 192.168.199.148
agent 192.168.199.148
```

There are a number of other useful output examples, such as `tcpdump`, output as NetFlow version 5 records, and a compact line-by-line output. This makes `sflowtool` very flexible to suit your monitoring environment.

ntop supports sFlow, which means you can directly export your sFlow to the ntop collector. If your collector is only NetFlow-aware, you can use the `-c` option for the `sflowtool` output in the NetFlow version 5 format:

```
$ sflowtool --help
...
tcpdump output:
   -t - (output in binary tcpdump(1) format)
   -r file - (read binary tcpdump(1) format)
   -x - (remove all IPV4 content)
   -z pad - (extend tcpdump pkthdr with this many zeros
                      e.g. try -z 8 for tcpdump on Red Hat Linux 6.2)

NetFlow output:
 -c hostname_or_IP - (netflow collector host)
 -d port - (netflow collector UDP port)
 -e - (netflow collector peer_as (default = origin_as))
 -s - (disable scaling of netflow output by sampling rate)
 -S - spoof source of netflow packets to input agent IP
```

Alternatively, you can also use InMon's sFlow-RT (http://www.sflow-rt.com/index.php) as your sFlow analytics engine. What sets sFlow-RT apart from an operator perspective is its vast REST API that can be customized to support your use cases. You can also easily retrieve the metrics from the API. You can take a look at its extensive API reference at http://www.sflow-rt.com/reference.php.

Note that sFlow-RT requires Java to run the following:

```
$ sudo apt-get install default-jre
$ java -version
openjdk version "1.8.0_121"
OpenJDK Runtime Environment (build 1.8.0_121-8u121-b13-0ubuntu1.16.04.2-b13)
OpenJDK 64-Bit Server VM (build 25.121-b13, mixed mode)
```

Once installed, downloading and running sFlow-RT is straightforward (https://sflow-rt.com/download.php):

```
$ wget http://www.inmon.com/products/sFlow-RT/sflow-rt.tar.gz
$ tar -xvzf sflow-rt.tar.gz
$ cd sflow-rt/
$ ./start.sh
```

```
2017-03-17T09:35:01-0700 INFO: Listening, sFlow port 6343
2017-03-17T09:35:02-0700 INFO: Listening, HTTP port 8008
```

We can point the web browser to HTTP port `8008` and verify the installation:

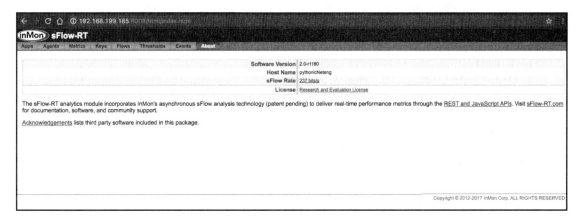

sFlow-RT about

As soon as sFlow-RT receives any sFlow packets, the agents and other metrics will appear:

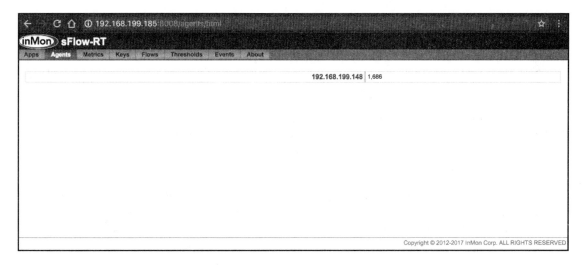

sFlow-RT agents

Here are two examples of using Python requests to retrieve information from sFlow-RT's REST API:

```
>>> import requests
>>> r = requests.get("http://192.168.199.185:8008/version")
>>> r.text
'2.0-r1180'
>>> r = requests.get("http://192.168.199.185:8008/agents/json")
>>> r.text
'{"192.168.199.148": {n "sFlowDatagramsLost": 0,n "sFlowDatagramSource":
["192.168.199.148"],n "firstSeen": 2195541,n "sFlowFlowDuplicateSamples":
0,n "sFlowDatagramsReceived": 441,n "sFlowCounterDatasources": 2,n
"sFlowFlowOutOfOrderSamples": 0,n "sFlowFlowSamples": 0,n
"sFlowDatagramsOutOfOrder": 0,n "uptime": 4060470520,n
"sFlowCounterDuplicateSamples": 0,n "lastSeen": 3631,n
"sFlowDatagramsDuplicates": 0,n "sFlowFlowDrops": 0,n
"sFlowFlowLostSamples": 0,n "sFlowCounterSamples": 438,n
"sFlowCounterLostSamples": 0,n "sFlowFlowDatasources": 0,n
"sFlowCounterOutOfOrderSamples": 0n}}'
```

Consult the reference documentation for additional REST endpoints available for your needs. Next, we will take a look at another tool called **Elasticsearch**, which is becoming pretty popular for both Syslog index and general network monitoring.

Elasticsearch (ELK stack)

As we have seen so far in this chapter, using just the Python tools as we have done would adequately monitor your network with enough scalability for all types of networks, large and small alike. However, I would like to introduce one additional open source, general-purpose, distributed search and analytics engine called **Elasticsearch** (https://www.elastic.co/). It is often referred to as just **Elastic** or **ELK stack** for combining **Elastic** with the frontend and input packages **Logstash**, and **Kibana**, respectively.

If you look at network monitoring in general, it is really about analyzing network data and making sense out of it. The ELK stack contains Elasticsearch, Logstash, and Kibana as a full stack to ingest information with Logstash, index and analyze data with Elasticsearch, and present the graphics output via Kibana. It is really three projects in one. It also has the flexibility to substitute Logstash with another input, such as **Beats**. Alternatively, you can use other tools, such as **Grafana**, instead of Kibana for visualization. The ELK stack by Elastic Co. also provides many add-on tools, referred to as **X-Pack**, for additional security, alerting, monitoring, and so on.

As you can probably tell by the description, ELK (or even Elasticsearch alone) is a deep topic to cover, and there are many books written on the subject. Even covering the basic usage would take up more space than we can spare in this book. I have considered leaving the subject out of the book simply because of its depth. However, ELK has become a very important tool for many of the projects that I am working on, including network monitoring. I feel leaving it out would be a huge disservice to you.

Therefore, I am going to take a few pages to briefly introduce the tool and a few use cases along with information for you to dig deeper if desired. We will go through the following topics:

- Setting up a hosted ELK service
- The Logstash format
- Python's helper script for Logstash formatting

Setting up a hosted ELK service

The entire ELK stack can be installed as a standalone server or distributed across multiple servers. The installation steps are available at `https://www.elastic.co/guide/en/elastic-stack/current/installing-elastic-stack.html`. In my experience, even with a minimal amount of data, a single VM running the ELK stack often stretches the resources. My first attempt at running ELK as a single VM lasted no more than a few days with barely two or three network devices sending log information toward it. After a few more unsuccessful attempts at running my own cluster as a beginner, I eventually settled on running the ELK stack as a hosted service, and this is what I would recommend you to start with.

As a hosted service, there are two providers that you can consider:

- **Amazon Elasticsearch Service** (`https://aws.amazon.com/elasticsearch-service/`)
- **Elastic Cloud** (`https://cloud.elastic.co/`)

Currently, AWS offers a free tier which is easy to get started with and is tightly integrated with the current suite of AWS tools, such as identity services (`https://aws.amazon.com/iam/`) and lambda functions (`https://aws.amazon.com/lambda/`). However, AWS's Elasticsearch Service does not have the latest features as compared to Elastic Cloud, nor does it have extended x-pack integration. However, because AWS offers a free tier, my recommendation would be that you start with the AWS Elasticsearch Service. If you find out later that you need more features than AWS can provide, you can always move to Elastic Cloud.

Setting up the service is straightforward; we just need to choose our region and give a name for our first domain. After setting it up, we can use the access policy to restrict input via an IP address; make sure this is the IP that AWS will see as the source IP (specify your corporate public IP if your host's IP address is translated behind the NAT firewall):

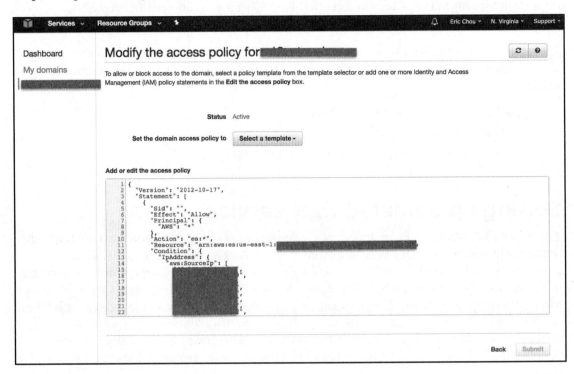

The Logstash format

Logstash can be installed on the server where you are comfortable sending your network log to. The installation steps are available at https://www.elastic.co/guide/en/logstash/current/installing-logstash.html. By default, you can put the Logstash configuration file under /etc/logstash/conf.d/. The file is in the input-filter-output format (https://www.elastic.co/guide/en/logstash/current/advanced-pipeline.html). In the following example, we specified the input as a network log file, with a placeholder for filtering the input, and the output as both printing out messages to the console as well as having the output exported toward our AWS Elasticsearch Service instance:

```
input {
    file {
```

```
      type => "network_log"
      path => "path to your network log file"
    }
  }
  filter {
    if [type] == "network_log" {
    }
  }
  output {
    stdout { codec => rubydebug }
    elasticsearch {
    index => "logstash_network_log-%{+YYYY.MM.dd}"
    hosts => ["http://<instance>.<region>.es.amazonaws.com"]
    }
  }
```

Now let's look at other things we can do with Python and Logstash.

Python helper script for Logstash formatting

The preceding Logstash configuration will allow us to ingest network logs and create the index on Elasticsearch. What would happen if the text format we intend on putting into ELK is not a standard log format? This is where Python can help. In the next example, we will perform the following:

1. Use the Python script to retrieve a list of IPs that the Spamhaus project considers to be a drop list (https://www.spamhaus.org/drop/drop.txt)
2. Use the Python logging module to format the information in such a way that Logstash can ingest it
3. Modify the Logstash configuration file so any new input could be sent to the AWS Elasticsearch Service

The chapter8_logstash_1.py script contains the code we will use. Besides the module imports, we will define the basic logging configuration. This section directly configures what the output would be and should be matched closely to the Logstash format:

```
#!/usr/env/bin python

#https://www.spamhaus.org/drop/drop.txt

import logging, pprint, re
import requests, json, datetime
from collections import OrderedDict
```

Network Monitoring with Python – Part 2

```
#logging configuration
logging.basicConfig(filename='./tmp/spamhaus_drop_list.log',
level=logging.INFO, format='%(asctime)s %(message)s', datefmt='%b %d
%I:%M:%S')
```

We will define a few more variables and save the list of IP addresses from the requests in a variable:

```
host = 'python_networking'
process = 'spamhause_drop_list'

r = requests.get('https://www.spamhaus.org/drop/drop.txt')
result = r.text.strip()

timeInUTC = datetime.datetime.utcnow().isoformat()
Item = OrderedDict()
Item["Time"] = timeInUTC
```

The final section of the script is a loop meant for parsing the output and writing it to the new log file:

```
for line in result.split('n'):
    if re.match('^;', line) or line == 'r': # comments
        next
    else:
        ip, record_number = line.split(";")
        logging.warning(host + ' ' + process + ': ' + 'src_ip=' +
ip.split("/")[0] + ' record_number=' + record_number.strip())
```

Here's a sample of the log file entry:

```
$ cat tmp/spamhaus_drop_list.log
...
Jul 14 11:35:26 python_networking spamhause_drop_list: src_ip=212.92.127.0
record_number=SBL352250
Jul 14 11:35:26 python_networking spamhause_drop_list: src_ip=216.47.96.0
record_number=SBL125132
Jul 14 11:35:26 python_networking spamhause_drop_list: src_ip=223.0.0.0
record_number=SBL230805
Jul 14 11:35:26 python_networking spamhause_drop_list: src_ip=223.169.0.0
record_number=SBL208009
...
```

We can then modify the Logstash configuration file accordingly to our new log format, starting with adding the input file location:

```
input {
  file {
```

```
      type => "network_log"
      path => "path to your network log file"
  }
   file {
      type => "spamhaus_drop_list"
      path =>
"/home/echou/Master_Python_Networking/Chapter8/tmp/spamhaus_drop_list.log"
   }
}
```

We can add more filter configurations using `grok`:

```
filter {
   if [type] == "spamhaus_drop_list" {
      grok {
         match => [ "message", "%{SYSLOGTIMESTAMP:timestamp}
%{SYSLOGHOST:hostname} %{NOTSPACE:process} src_ip=%{IP:src_ip}
%{NOTSPACE:record_number}.*"]
         add_tag => ["spamhaus_drop_list"]
      }
   }
}
```

We can leave the output section unchanged, as the additional entries will be stored in the same index. We can now use the ELK stack to query, store, and view the network log as well as the Spamhaus IP information.

Summary

In this chapter, we looked at additional ways in which we can utilize Python to enhance our network monitoring effort. We began by using Python's Graphviz package to create network topology graphs with real-time LLDP information reported by the network devices. This allows us to effortlessly show the current network topology as well as easily notice any link failures.

Next, we used Python to parse NetFlow version 5 packets to enhance our understanding and troubleshooting of NetFlow. We also looked at how to use ntop and Python to extend ntop for NetFlow monitoring. sFlow is an alternative packet sampling technology that we looked at where we use `sflowtool` and sFlow-RT to interpret the results. We ended the chapter with a general-purpose data analyzing tool, namely Elasticsearch, or the ELK stack.

In Chapter 9, *Building Network Web Services with Python*, we will explore how to use the Python web framework Flask to build network web services.

9
Building Network Web Services with Python

In the previous chapters, we were a consumer of the APIs provided by various tools. In Chapter 3, *APIs and Intent-Driven Networking*, we saw that we can use a HTTP POST method to NX-API at the `http://<your router ip>/ins` URL with the CLI command embedded in the body to execute commands remotely on the Cisco Nexus device; the device then returns the command execution output in return. In Chapter 8, *Network Monitoring with Python – Part 2*, we used the GET method for our sFlow-RT at `http://<your host ip>:8008/version` with an empty body to retrieve the version of the sFlow-RT software. These exchanges are examples of RESTful web services.

According to Wikipedia (https://en.wikipedia.org/wiki/Representational_state_transfer):

> *"Representational state transfer (REST) or RESTful web services is one way of providing interoperability between computer systems on the internet. REST-compliant web services allow requesting systems to access and manipulate the textual representation of web resources using a uniform and predefined set of stateless operations."*

As noted, REST web services using the HTTP protocol is only one of many methods of information exchange on the web; other forms of web services also exist. However, it is the most commonly used web service today, with the associated GET, POST, PUT, and DELETE verbs as a predefined way of information exchange.

One of the advantages of using RESTful services is the ability it provides for you to hide your internal operations from the user while still providing them with the service. For example, in the case of sFlow-RT, if we were to log in to the device where our software is installed, we would need more in-depth knowledge of the tool to know where to check for the software version. However, by providing the resources in the form of a URL, the software abstracts the version-checking operations from the requester, making the operation much simpler. The abstraction also provides a layer of security, as it can now open up the endpoints only as needed.

As the master of the network universe, RESTful web services provide many notable benefits that we can enjoy, such as the following:

- You can abstract the requester from learning about the internals of the network operations. For example, we can provide a web service to query the switch version without the requester having to know the exact CLI command or API format required.
- We can consolidate and customize operations that uniquely fit our network needs, such as a resource to upgrade all our top-of-rack switches.
- We can provide better security by only exposing the operations as needed. For example, we can provide read-only URLs (GET) to core network devices and read-write URLs (GET / POST / PUT / DELETE) to access-level switches.

In this chapter, we will use one of the most popular Python web frameworks, **Flask**, to create our own REST web service for our network. In this chapter, we will learn about the following:

- Comparing Python web frameworks
- Introduction to Flask
- Operations involving static network contents
- Operations involving dynamic network operations

Let's get started by looking at the available Python web frameworks and why we chose Flask.

Comparing Python web frameworks

Python is known for its great many web frameworks. There is a running joke at PyCon, which is that you can never work as a full-time Python developer without working with any of the Python web frameworks. There is even an annual conference held for Django, one of the most popular Python frameworks, called DjangoCon. It attracts hundreds of attendees every year. If you sort the Python web frameworks on `https://hotframeworks.com/languages/python`, you can see that there is no shortage of choices when it comes to Python and web frameworks:

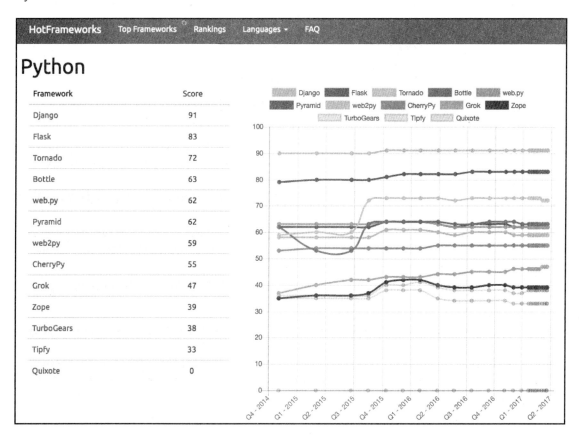

Python web frameworks ranking

With so many options to choose from, which framework should we pick? Clearly, trying all the frameworks out yourself would be really time-consuming. The question about which web framework is better is also a passionate topic among web developers. If you ask this question on any of the forums, such as Quora, or search on Reddit, get ready for some highly opinionated answers and heated debates.

Speaking of Quora and Reddit, here's an interesting fact: both Quora and Reddit were written in Python. Reddit uses Pylons (https://www.reddit.com/wiki/faq#wiki_so_what_python_framework_do_you_use.3F), while Quora started with Pylons but replaced a portion of the framework with their in-house code (https://www.quora.com/What-languages-and-frameworks-are-used-to-code-Quora).

Of course, I have my own bias toward programming languages (Python!) and web frameworks (Flask!). In this section, I hope to convey to you my reasoning behind choosing one over the other. Let's pick the top two frameworks from the preceding HotFrameworks list and compare them:

- **Django**: The self-proclaimed "web framework for perfectionists with deadlines" is a high-level Python web framework that encourages rapid development and a clean, pragmatic design (https://www.djangoproject.com/). It is a large framework with pre-built code that provides an administrative panel and built-in content management.
- **Flask**: This is a microframework for Python and is based on Werkzeug, Jinja2, and good intentions (http://flask.pocoo.org/). By being a microframework, Flask intends on keeping the core small and being easy to extend when needed. The "micro" in microframework does not mean that Flask is lacking in functionality, nor does it mean it cannot work in a production environment.

Personally, I find Django a bit difficult to extend, and most of the time, I only use a fraction of the pre-built code. The Django framework also has a strong opinion on how things should be done; any deviation from it would sometimes leave the user feeling that they are "fighting with the framework". For example, if you look at the Django Database documentation, (https://docs.djangoproject.com/en/2.1/ref/databases/), you will notice that the framework supports a number of different SQL databases. However, they are all variants of SQL database such as MySQL, PostgreSQL, SQLite, and others. What if you want to use a NoSQL database such as MongoDB or CouchDB? It might be possible but could be leaving you in your own hands. Being an opinionated framework is certainly not a bad thing, it is just a matter of opinion (no pun intended).

The idea of keeping the core code small and extending it when needed is very appealing to me. The initial example on the documentation to get Flask up and running consists of only eight lines of code and is easy to understand, even if you don't have any prior experience. Since Flask is built with extensions in mind, writing your own extensions, such as decorator, is pretty easy. Even though it is a microframework, the Flask core still includes the necessary components, such as a development server, debugger, integration with unit tests, RESTful request dispatching, and more, to get you started out of the box. As you can see, besides Django, Flask is the second most popular Python framework by some measure. The popularity that comes with community contribution, support, and quick development helps it further expand its reach.

For the preceding reasons, I feel that Flask is an ideal choice for us when it comes to building network web services.

Flask and lab setup

In this chapter, we will use `virtualenv` to isolate the environment we will work in. As the name indicates, virtualenv is a tool that creates a virtual environment. It can keep the dependencies required by different projects in separate places while keeping the global site-packages clean. In other words, when you install Flask in the virtual environment, it is only installed in the local `virtualenv` project directory, not the global site-packages. This make porting the code to other places very easy.

The chances are high that you may have already come across `virtualenv` while working with Python before, so we will run through this process quickly. If you have not, feel free to pick up one of many excellent tutorials online, such as http://docs.python-guide.org/en/latest/dev/virtualenvs/.

To use , we will first need to install `virtualenv`:

```
# Python 3
$ sudo apt-get install python3-venv
$ python3 -m venv venv

# Python 2
$ sudo apt-get install python-virtualenv
$ virtualenv venv-python2
```

The proceeding command uses the `venv` module (`-m venv`) to get a `venv` folder with a full Python interpreter inside it. We can use `source venv/bin/activate` and `deactivate` to move in and out of the local Python environment:

```
$ source venv/bin/activate
(venv) $ python
$ which python
/home/echou/Master_Python_Networking_second_edition/Chapter09/venv/bin/pyth
on
$ python
Python 3.5.2 (default, Nov 23 2017, 16:37:01)
[GCC 5.4.0 20160609] on linux
Type "help", "copyright", "credits" or "license" for more information.
>>>
>>> exit()
(venv) $ deactivate
```

In this chapter, we will install quite a few Python packages. To make life easier, I have included a `requirements.txt` file on the book's GitHub repository; we can use it to install all the necessary packages (remember to activate your virtualenv). You should see packages being downloaded and successfully installed at the end of the process:

```
(venv) $ pip install -r requirements.txt
Collecting Flask==0.10.1 (from -r requirements.txt (line 1))
  Downloading
https://files.pythonhosted.org/packages/db/9c/149ba60c47d107f85fe5256413334
8458f093dd5e6b57a5b60ab9ac517bb/Flask-0.10.1.tar.gz (544kB)
    100% |████████████████████████████████| 552kB
2.0MB/s
Collecting Flask-HTTPAuth==2.2.1 (from -r requirements.txt (line 2))
  Downloading
https://files.pythonhosted.org/packages/13/f3/efc053c66a7231a5a38078a813aee
06cd63ca90ab1b3e269b63edd5ff1b2/Flask-HTTPAuth-2.2.1.tar.gz
... <skip>
  Running setup.py install for Pygments ... done
  Running setup.py install for python-dateutil ... done
Successfully installed Flask-0.10.1 Flask-HTTPAuth-2.2.1 Flask-
SQLAlchemy-1.0 Jinja2-2.7.3 MarkupSafe-0.23 Pygments-1.6 SQLalchemy-0.9.6
Werkzeug-0.9.6 httpie-0.8.0 itsdangerous-0.24 python-dateutil-2.2
requests-2.3.0 six-1.11.0
```

For our network topology, we will use a simple four-node network, as shown here:

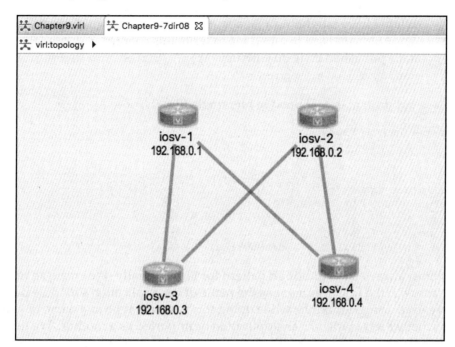

Lab topology

Let's take a look at Flask in the next section.

 Please note that, from here on out, I will assume that you will always execute from the virtual environment and that you have installed the necessary packages in the `requirements.txt` file.

Introduction to Flask

Like most popular open source projects, Flask has very good documentation, which is available at `http://flask.pocoo.org/docs/0.10/`. If any of the examples are unclear, you can be sure to find the answer on the project documentation.

I would also highly recommend Miguel Grinberg's (https://blog.miguelgrinberg.com/) work related to Flask. His blog, book, and video training have taught me a lot about Flask. In fact, Miguel's class *Building Web APIs with Flask* inspired me to write this chapter. You can take a look at his published code on GitHub: https://github.com/miguelgrinberg/oreilly-flask-apis-video.

Our first Flask application is contained in one single file, `chapter9_1.py`:

```
from flask import Flask
app = Flask(__name__)

@app.route('/')
def hello_networkers():
    return 'Hello Networkers!'

if __name__ == '__main__':
    app.run(host='0.0.0.0', debug=True)
```

This will almost always be your design pattern for Flask initially. We create an instance of the Flask class with the first argument as the name of the application's module package. In this case, we used a single module; while doing this yourself, type in a name of your choice to indicate whether it is started as an application or imported as a module. We then use the route decorator to tell Flask which URL should be handled by the `hello_networkers()` function; in this case, we indicated the root path. We end the file with the usual name (https://docs.python.org/3.5/library/__main__.html). We only added the host and debug options, which allow more verbose output and also allow us to listen on all the interfaces of the host (by default, it only listens on loopback). We can run this application using the development server:

```
(venv) $ python chapter9_1.py
 * Running on http://0.0.0.0:5000/
 * Restarting with reloader
```

Now that we have a server running, let's test the server response with an HTTP client.

The HTTPie client

We have already installed HTTPie (https://httpie.org/) as part of the installation from reading the requirements.txt file. Although this book is printed in black and white text so it does not show up here, in your installation, you can see that HTTPie has better syntax highlighting for HTTP transactions. It also has a more intuitive command-line interaction with the RESTful HTTP server. We can use it to test our first Flask application (more examples on HTTPie to follow):

```
$ http GET http://172.16.1.173:5000/
HTTP/1.0 200 OK
Content-Length: 17
Content-Type: text/html; charset=utf-8
Date: Wed, 22 Mar 2017 17:37:12 GMT
Server: Werkzeug/0.9.6 Python/3.5.2

Hello Networkers!
```

Alternatively, you can also use the -i switch with curl to see the HTTP headers: curl -i http://172.16.1.173:5000/.

We will use HTTPie as our client for this chapter; it is worth taking a minute or two to take a look at its usage. We will use the free website HTTP Bin (https://httpbin.org/) to show the use of HTTPie. The usage of HTTPie follows this simple pattern:

```
$ http [flags] [METHOD] URL [ITEM]
```

Following the preceding pattern, a GET request is very straightforward, as we have seen with our Flask development server:

```
$ http GET https://httpbin.org/user-agent
...
{
 "user-agent": "HTTPie/0.8.0"
}
```

JSON is the default implicit content type for `HTTPie`. If your HTTP body contains just strings, no other operation is needed. If you need to apply non-string JSON fields, use `:=` or other documented special characters:

```
$ http POST https://httpbin.org/post name=eric twitter=at_ericchou
married:=true
HTTP/1.1 200 OK
...
Content-Type: application/json
...
{
    "headers": {
...
        "User-Agent": "HTTPie/0.8.0"
    },
    "json": {
        "married": true,
        "name": "eric",
        "twitter": "at_ericchou"
    },
    ...
    "url": "https://httpbin.org/post"
}
```

As you can see, `HTTPie` is a big improvement from the traditional curl syntax and makes testing the REST API a breeze.

More usage examples are available at `https://httpie.org/doc#usage`.

Getting back to our Flask program, a large part of API building is based on the flow of URL routing. Let's take a deeper look at the `app.route()` decorator.

URL routing

We added two additional functions and paired them up with the appropriate `app.route()` route in `chapter9_2.py`:

```
$ cat chapter9_2.py
from flask import Flask
app = Flask(__name__)

@app.route('/')
```

```
def index():
    return 'You are at index()'

@app.route('/routers/')
def routers():
    return 'You are at routers()'

if __name__ == '__main__':
    app.run(host='0.0.0.0', debug=True)
```

The result is that different endpoints are passed to different functions. We can verify this with two `http` requests:

```
# Server
$ python chapter9_2.py

# Client
$ http GET http://172.16.1.173:5000/
...

You are at index()

$ http GET http://172.16.1.173:5000/routers/
...

You are at routers()
```

Of course, the routing will be pretty limited if we have to keep it static all the time. There are ways to pass variables from the URL to Flask; we will look at an example of this in the coming section.

URL variables

As mentioned previously, we can also pass variables to the URL, as seen in the examples discussed in `chapter9_3.py`:

```
...
@app.route('/routers/<hostname>')
def router(hostname):
    return 'You are at %s' % hostname

@app.route('/routers/<hostname>/interface/<int:interface_number>')
def interface(hostname, interface_number):
    return 'You are at %s interface %d' % (hostname, interface_number)
...
```

Note that, in the `/routers/<hostname>` URL, we pass the `<hostname>` variable as a string; `<int:interface_number>` will specify that the variable should only be an integer:

```
$ http GET http://172.16.1.173:5000/routers/host1
...
You are at host1

$ http GET http://172.16.1.173:5000/routers/host1/interface/1
...
You are at host1 interface 1

# Throws exception
$ http GET http://172.16.1.173:5000/routers/host1/interface/one
HTTP/1.0 404 NOT FOUND
...
<!DOCTYPE HTML PUBLIC "-//W3C//DTD HTML 3.2 Final//EN">
<title>404 Not Found</title>
<h1>Not Found</h1>
<p>The requested URL was not found on the server. If you entered the URL manually please check your spelling and try again.</p>
```

The converter includes integers, float, and path (it accepts slashes).

Besides matching static routes, we can also generate URLs on the fly. This is very useful when we do not know the endpoint variable in advance or if the endpoint is based on other conditions, such as the values queried from a database. Let's take a look at an example of this.

URL generation

In `chapter9_4.py`, we wanted to dynamically create a URL in the form of `'/<hostname>/list_interfaces'` in code:

```
from flask import Flask, url_for
...
@app.route('/<hostname>/list_interfaces')
def device(hostname):
    if hostname in routers:
        return 'Listing interfaces for %s' % hostname
    else:
        return 'Invalid hostname'

routers = ['r1', 'r2', 'r3']
for router in routers:
    with app.test_request_context():
```

```
            print(url_for('device', hostname=router))
...
```

Upon its execution, you will have a nice and logical URL, as follows:

```
(venv) $ python chapter9_4.py
/r1/list_interfaces
/r2/list_interfaces
/r3/list_interfaces
 * Running on http://0.0.0.0:5000/
 * Restarting with reloader
```

For now, you can think of `app.text_request_context()` as a dummy `request` object that is necessary for demonstrative purposes. If you are interested in the local context, feel free to take a look at http://werkzeug.pocoo.org/docs/0.14/local/.

The jsonify return

Another time saver in Flask is the `jsonify()` return, which wraps `json.dumps()` and turns the JSON output into a `response` object with `application/json` as the content type in the HTTP header. We can tweak the last script a bit, just like we will do in `chapter9_5.py`:

```
from flask import Flask, jsonify

app = Flask(__name__)

@app.route('/routers/<hostname>/interface/<int:interface_number>')
def interface(hostname, interface_number):
    return jsonify(name=hostname, interface=interface_number)

if __name__ == '__main__':
    app.run(host='0.0.0.0', debug=True)
```

We will see the result returned as a JSON object with the appropriate header:

```
$ http GET http://172.16.1.173:5000/routers/r1/interface/1
HTTP/1.0 200 OK
Content-Length: 36
Content-Type: application/json
...

{
    "interface": 1,
    "name": "r1"
}
```

Having looked at URL routing and the `jsonify()` return in Flask, we are now ready to build an API for our network.

Network resource API

Often, your network consists of network devices that do not change a lot once put into production. For example, you would have core devices, distribution devices, spine, leaf, top-of-rack switches, and so on. Each of the devices would have certain characteristics and features that you would like to keep in a persistent location so that you can easily retrieve them later on. This is often done in terms of storing data in a database. However, you would not normally want to give other users, who might want this information, direct access to the database; nor do they want to learn all the complex SQL query language. For this case, we can leverage Flask and the **Flask-SQLAlchemy** extension of Flask.

> You can learn more about Flask-SQLAlchemy at `http://flask-sqlalchemy.pocoo.org/2.1/`.

Flask-SQLAlchemy

Of course, SQLAlchemy and the Flask extension are a database abstraction layer and object relational mapper, respectively. It's a fancy way of saying to use the `Python` object for a database. To make things simple, we will use SQLite as the database, which is a flat file that acts as a self-contained SQL database. We will look at the content of `chapter9_db_1.py` as an example of using Flask-SQLAlchemy to create a network database and insert a table entry into the database.

To begin with, we will create a Flask application and load the configuration for SQLAlchemy, such as the database path and name, then create the `SQLAlchemy` object by passing the application to it:

```
from flask import Flask
from flask_sqlalchemy import SQLAlchemy

# Create Flask application, load configuration, and create
# the SQLAlchemy object
app = Flask(__name__)
app.config['SQLALCHEMY_DATABASE_URI'] = 'sqlite:///network.db'
db = SQLAlchemy(app)
```

We can then create a database object and its associated primary key and various columns:

```
class Device(db.Model):
    __tablename__ = 'devices'
    id = db.Column(db.Integer, primary_key=True)
    hostname = db.Column(db.String(120), index=True)
    vendor = db.Column(db.String(40))

    def __init__(self, hostname, vendor):
        self.hostname = hostname
        self.vendor = vendor

    def __repr__(self):
        return '<Device %r>' % self.hostname
```

We can invoke the database object, create entries, and insert them into the database table. Keep in mind that anything we add to the session needs to be committed to the database in order to be permanent:

```
if __name__ == '__main__':
    db.create_all()
    r1 = Device('lax-dc1-core1', 'Juniper')
    r2 = Device('sfo-dc1-core1', 'Cisco')
    db.session.add(r1)
    db.session.add(r2)
    db.session.commit()
```

We will run the Python script and check for the existence of the database file:

```
$ python chapter9_db_1.py
$ ls network.db
network.db
```

We can use the interactive prompt to check the database table entries:

```
>>> from flask import Flask
>>> from flask_sqlalchemy import SQLAlchemy
>>>
>>> app = Flask(__name__)
>>> app.config['SQLALCHEMY_DATABASE_URI'] = 'sqlite:///network.db'
>>> db = SQLAlchemy(app)
>>> from chapter9_db_1 import Device
>>> Device.query.all()
[<Device 'lax-dc1-core1'>, <Device 'sfo-dc1-core1'>]
>>> Device.query.filter_by(hostname='sfo-dc1-core1')
<flask_sqlalchemy.BaseQuery object at 0x7f1b4ae07eb8>
>>> Device.query.filter_by(hostname='sfo-dc1-core1').first()
<Device 'sfo-dc1-core1'>
```

We can also create new entries in the same manner:

```
>>> r3 = Device('lax-dc1-core2', 'Juniper')
>>> db.session.add(r3)
>>> db.session.commit()
>>> Device.query.all()
[<Device 'lax-dc1-core1'>, <Device 'sfo-dc1-core1'>, <Device 'lax-dc1-core2'>]
```

Network content API

Before we dive into the code, let's take a moment to think about the API that we are trying to create. Planning for an API is usually more art than science; it really depends on your situation and preference. What I suggest next is, by no means, the right way, but for now, stay with me for the purposes of getting started.

Recall that, in our diagram, we have four Cisco IOSv devices. Let's pretend that two of them, `iosv-1` and `iosv-2`, are of the network role of the spine. The other two devices, `iosv-3` and `iosv-4`, are in our network service as leafs. These are obviously arbitrary choices and can be modified later on, but the point is that we want to serve data about our network devices and expose them via an API.

To make things simple, we will create two APIs: a devices group API and a single device API:

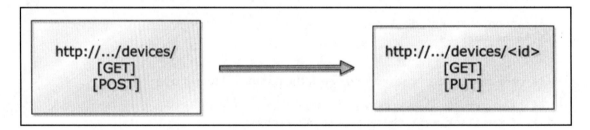

Network content API

The first API will be our `http://172.16.1.173/devices/` endpoint that supports two methods: `GET` and `POST`. The `GET` request will return the current list of devices, while the `POST` request with the proper JSON body will create the device. Of course, you can choose to have different endpoints for creation and query, but in this design, we choose to differentiate the two by the HTTP methods.

The second API will be specific to our device in the form of
`http://172.16.1.173/devices/<device id>`. The API with the GET request will show
the details of the device that we have entered into the database. The PUT request will
modify the entry with the update. Note that we use PUT instead of POST. This is typical of
HTTP API usage; when we need to modify an existing entry, we will use PUT instead of
POST.

At this point, you should have a good idea about what your API will look like. To better
visualize the end result, I am going to jump ahead and show the end result quickly before
we take a look at the code.

A POST request to the `/devices/` API will allow you to create an entry. In this case, I
would like to create our network device with attributes such as hostname, loopback IP,
management IP, role, vendor, and the operating system it runs on:

```
$ http POST http://172.16.1.173:5000/devices/ 'hostname'='iosv-1'
'loopback'='192.168.0.1' 'mgmt_ip'='172.16.1.225' 'role'='spine'
'vendor'='Cisco' 'os'='15.6'
HTTP/1.0 201 CREATED
Content-Length: 2
Content-Type: application/json
Date: Fri, 24 Mar 2017 01:45:15 GMT
Location: http://172.16.1.173:5000/devices/1
Server: Werkzeug/0.9.6 Python/3.5.2

{}
```

I can repeat the preceding step for the additional three devices:

```
$ http POST http://172.16.1.173:5000/devices/ 'hostname'='iosv-2'
'loopback'='192.168.0.2' 'mgmt_ip'='172.16.1.226' 'role'='spine'
'vendor'='Cisco' 'os'='15.6'
...
$ http POST http://172.16.1.173:5000/devices/ 'hostname'='iosv-3',
'loopback'='192.168.0.3' 'mgmt_ip'='172.16.1.227' 'role'='leaf'
'vendor'='Cisco' 'os'='15.6'
...
$ http POST http://172.16.1.173:5000/devices/ 'hostname'='iosv-4',
'loopback'='192.168.0.4' 'mgmt_ip'='172.16.1.228' 'role'='leaf'
'vendor'='Cisco' 'os'='15.6'
```

If we can use the same API with the GET request, we will be able to see the list of network
devices that we created:

```
$ http GET http://172.16.1.173:5000/devices/
HTTP/1.0 200 OK
```

```
Content-Length: 188
Content-Type: application/json
Date: Fri, 24 Mar 2017 01:53:15 GMT
Server: Werkzeug/0.9.6 Python/3.5.2

{
    "device": [
        "http://172.16.1.173:5000/devices/1",
        "http://172.16.1.173:5000/devices/2",
        "http://172.16.1.173:5000/devices/3",
        "http://172.16.1.173:5000/devices/4"
    ]
}
```

Similarly, using the GET request for /devices/<id> will return specific information related to the device:

```
$ http GET http://172.16.1.173:5000/devices/1
HTTP/1.0 200 OK
Content-Length: 188
Content-Type: application/json
...
{
    "hostname": "iosv-1",
    "loopback": "192.168.0.1",
    "mgmt_ip": "172.16.1.225",
    "os": "15.6",
    "role": "spine",
    "self_url": "http://172.16.1.173:5000/devices/1",
    "vendor": "Cisco"
}
```

Let's pretend we have downgraded the r1 operating system from 15.6 to 14.6. We can use the PUT request to update the device record:

```
$ http PUT http://172.16.1.173:5000/devices/1 'hostname'='iosv-1'
'loopback'='192.168.0.1' 'mgmt_ip'='172.16.1.225' 'role'='spine'
'vendor'='Cisco' 'os'='14.6'
HTTP/1.0 200 OK

# Verification
$ http GET http://172.16.1.173:5000/devices/1
...
{
    "hostname": "r1",
    "loopback": "192.168.0.1",
    "mgmt_ip": "172.16.1.225",
    "os": "14.6",
```

```
    "role": "spine",
    "self_url": "http://172.16.1.173:5000/devices/1",
    "vendor": "Cisco"
}
```

Now, let's take a look at the code in chapter9_6.py that helped create the preceding APIs. What's cool, in my opinion, is that all of these APIs were done in a single file, including the database interaction. Later on, when we outgrow the APIs at hand, we can always separate the components out, such as having a separate file for the database class.

Devices API

The chapter9_6.py file starts with the necessary imports. Note that the following request import is the request object from the client and not the requests package that we were using in the previous chapters:

```
from flask import Flask, url_for, jsonify, request
from flask_sqlalchemy import SQLAlchemy
# The following is deprecated but still used in some examples
# from flask.ext.sqlalchemy import SQLAlchemy
```

We declared a database object with its id as the primary key and string fields for hostname, loopback, mgmt_ip, role, vendor, and os:

```
class Device(db.Model):
    __tablename__ = 'devices'
    id = db.Column(db.Integer, primary_key=True)
    hostname = db.Column(db.String(64), unique=True)
    loopback = db.Column(db.String(120), unique=True)
    mgmt_ip = db.Column(db.String(120), unique=True)
    role = db.Column(db.String(64))
    vendor = db.Column(db.String(64))
    os = db.Column(db.String(64))
```

The get_url() function returns a URL from the url_for() function. Note that the get_device() function that's called is not defined just yet under the '/devices/<int:id>' route:

```
def get_url(self):
    return url_for('get_device', id=self.id, _external=True)
```

The `export_data()` and `import_data()` functions are mirror images of each other. One is used to get the information from the database to the user (`export_data()`) when we use the `GET` method. The other is to put information from the user to the database (`import_data()`) when we use the `POST` or `PUT` method:

```python
def export_data(self):
    return {
        'self_url': self.get_url(),
        'hostname': self.hostname,
        'loopback': self.loopback,
        'mgmt_ip': self.mgmt_ip,
        'role': self.role,
        'vendor': self.vendor,
        'os': self.os
    }

def import_data(self, data):
    try:
        self.hostname = data['hostname']
        self.loopback = data['loopback']
        self.mgmt_ip = data['mgmt_ip']
        self.role = data['role']
        self.vendor = data['vendor']
        self.os = data['os']
    except KeyError as e:
        raise ValidationError('Invalid device: missing ' + e.args[0])
    return self
```

With the `database` object in place as well as the import and export functions created, the URL dispatch is straightforward for device operations. The `GET` request will return a list of devices by querying all the entries in the devices table and also return the URL of each entry. The `POST` method will use the `import_data()` function with the global `request` object as the input. It will then add the device and commit the information to the database:

```python
@app.route('/devices/', methods=['GET'])
def get_devices():
    return jsonify({'device': [device.get_url()
                               for device in Device.query.all()]})

@app.route('/devices/', methods=['POST'])
def new_device():
    device = Device()
    device.import_data(request.json)
    db.session.add(device)
    db.session.commit()
    return jsonify({}), 201, {'Location': device.get_url()}
```

If you look at the POST method, the returned body is an empty JSON body, with the status code 201 (created) as well as extra headers:

```
HTTP/1.0 201 CREATED
Content-Length: 2
Content-Type: application/json
Date: ...
Location: http://172.16.1.173:5000/devices/4
Server: Werkzeug/0.9.6 Python/3.5.2
```

Let's look at the API that queries and returns information pertaining to individual devices.

The device ID API

The route for individual devices specifies that the ID should be an integer, which can act as our first line of defense against a bad request. The two endpoints follow the same design pattern as our /devices/ endpoint, where we use the same import and export functions:

```
@app.route('/devices/<int:id>', methods=['GET'])
def get_device(id):
    return jsonify(Device.query.get_or_404(id).export_data())

@app.route('/devices/<int:id>', methods=['PUT'])
def edit_device(id):
    device = Device.query.get_or_404(id)
    device.import_data(request.json)
    db.session.add(device)
    db.session.commit()
    return jsonify({})
```

Note the query_or_404() method; it provides a convenient way for returning 404 (not found) if the database query returns negative for the ID passed in. This is a pretty elegant way of providing a quick check on the database query.

Finally, the last part of the code creates the database table and starts the Flask development server:

```
if __name__ == '__main__':
    db.create_all()
    app.run(host='0.0.0.0', debug=True)
```

This is one of the longer Python scripts in this book, which is why we took more time to explain it in detail. The script provides a way to illustrate how we can utilize the database in the backend to keep track of the network devices and only expose them to the external world as APIs, using Flask.

In the next section, we will take a look at how to use the API to perform asynchronous tasks on either individual devices or a group of devices.

Network dynamic operations

Our API can now provide static information about the network; anything that we can store in the database can be returned to the requester. It would be great if we can interact with our network directly, such as a query for the device information or to push configuration changes to the device.

We will start this process by leveraging the script we have already seen in Chapter 2, *Low-Level Network Device Interactions*, for interacting with a device via Pexpect. We will modify the script slightly into a function we can repeatedly use in chapter9_pexpect_1.py:

```
# We need to install pexpect for our virtual env
$ pip install pexpect

$ cat chapter9_pexpect_1.py
import pexpect

def show_version(device, prompt, ip, username, password):
    device_prompt = prompt
    child = pexpect.spawn('telnet ' + ip)
    child.expect('Username:')
    child.sendline(username)
    child.expect('Password:')
    child.sendline(password)
    child.expect(device_prompt)
    child.sendline('show version | i V')
    child.expect(device_prompt)
    result = child.before
    child.sendline('exit')
    return device, result
```

We can test the new function via the interactive prompt:

```
$ pip3 install pexpect
$ python
>>> from chapter9_pexpect_1 import show_version
>>> print(show_version('iosv-1', 'iosv-1#', '172.16.1.225', 'cisco', 'cisco'))
('iosv-1', b'show version | i V\r\nCisco IOS Software, IOSv Software (VIOS-ADVENTERPRISEK9-M), Version 15.6(3)M2, RELEASE SOFTWARE (fc2)\r\n')
>>>
```

Chapter 9

 Make sure that your Pexpect script works before you proceed. The following code assumes that you have entered the necessary database information from the previous section.

We can add a new API for querying the device version in chapter9_7.py:

```
from chapter9_pexpect_1 import show_version
...
@app.route('/devices/<int:id>/version', methods=['GET'])
def get_device_version(id):
    device = Device.query.get_or_404(id)
    hostname = device.hostname
    ip = device.mgmt_ip
    prompt = hostname+"#"
    result = show_version(hostname, prompt, ip, 'cisco', 'cisco')
    return jsonify({"version": str(result)})
```

The result will be returned to the requester:

```
$ http GET http://172.16.1.173:5000/devices/4/version
HTTP/1.0 200 OK
Content-Length: 210
Content-Type: application/json
Date: Fri, 24 Mar 2017 17:05:13 GMT
Server: Werkzeug/0.9.6 Python/3.5.2

{
  "version": "('iosv-4', b'show version | i V\r\nCisco IOS Software, IOSv Software (VIOS-ADVENTERPRISEK9-M), Version 15.6(2)T, RELEASE SOFTWARE (fc2)\r\nProcessor board ID 9U96V39A4Z12PCG4O6Y0Q\r\n')"
}
```

We can also add another endpoint that will allow us to perform a bulk action on multiple devices, based on their common fields. In the following example, the endpoint will take the device_role attribute in the URL and match it up with the appropriate device(s):

```
@app.route('/devices/<device_role>/version', methods=['GET'])
def get_role_version(device_role):
    device_id_list = [device.id for device in Device.query.all() if device.role == device_role]
    result = {}
    for id in device_id_list:
        device = Device.query.get_or_404(id)
        hostname = device.hostname
        ip = device.mgmt_ip
        prompt = hostname + "#"
```

[307]

```
            device_result = show_version(hostname, prompt, ip, 'cisco',
'cisco')
            result[hostname] = str(device_result)
    return jsonify(result)
```

 Of course, looping through all the devices in `Device.query.all()` is not efficient, as in the preceding code. In production, we will use a SQL query that specifically targets the role of the device.

When we use the REST API, we can see that all the spine, as well as leaf, devices can be queried at the same time:

```
$ http GET http://172.16.1.173:5000/devices/spine/version
HTTP/1.0 200 OK
...
{
 "iosv-1": "('iosv-1', b'show version | i V\r\nCisco IOS Software, IOSv
Software (VIOS-ADVENTERPRISEK9-M), Version 15.6(2)T, RELEASE SOFTWARE
(fc2)\r\n')",
 "iosv-2": "('iosv-2', b'show version | i V\r\nCisco IOS Software, IOSv
Software (VIOS-ADVENTERPRISEK9-M), Version 15.6(2)T, RELEASE SOFTWARE
(fc2)\r\nProcessor board ID 9T7CB2J2V6F0DLWK7V48E\r\n')"
}


$ http GET http://172.16.1.173:5000/devices/leaf/version
HTTP/1.0 200 OK
...
{
 "iosv-3": "('iosv-3', b'show version | i V\r\nCisco IOS Software, IOSv
Software (VIOS-ADVENTERPRISEK9-M), Version 15.6(2)T, RELEASE SOFTWARE
(fc2)\r\nProcessor board ID 9MGG8EA1E0V2PE2D8KDD7\r\n')",
 "iosv-4": "('iosv-4', b'show version | i V\r\nCisco IOS Software, IOSv
Software (VIOS-ADVENTERPRISEK9-M), Version 15.6(2)T, RELEASE SOFTWARE
(fc2)\r\nProcessor board ID 9U96V39A4Z12PCG4O6Y0Q\r\n')"
}
```

As illustrated, the new API endpoints query the device(s) in real time and return the result to the requester. This works relatively well when you can guarantee a response from the operation within the timeout value of the transaction (30 seconds, by default) or if you are OK with the HTTP session timing out before the operation is completed. One way to deal with the timeout issue is to perform the tasks asynchronously. We will look at how to do so in the next section.

Asynchronous operations

Asynchronous operations are, in my opinion, an advanced topic of Flask. Luckily, Miguel Grinberg (https://blog.miguelgrinberg.com/), whose Flask work I am a big fan of, provides many posts and examples on his blog and on GitHub. For asynchronous operations, the example code in `chapter9_8.py` referenced Miguel's GitHub code on the Raspberry Pi file (https://github.com/miguelgrinberg/oreilly-flask-apis-video/blob/master/camera/camera.py) for the background decorator. We will start by importing a few more modules:

```
from flask import Flask, url_for, jsonify, request,
    make_response, copy_current_request_context
...
import uuid
import functools
from threading import Thread
```

The background decorator takes in a function and runs it as a background task using thread and UUID for the task ID. It returns the status code `202` accepted and the location of the new resources for the requester to check. We will make a new URL for status checking:

```
@app.route('/status/<id>', methods=['GET'])
def get_task_status(id):
    global background_tasks
    rv = background_tasks.get(id)
    if rv is None:
        return not_found(None)

    if isinstance(rv, Thread):
        return jsonify({}), 202, {'Location': url_for('get_task_status', id=id)}

    if app.config['AUTO_DELETE_BG_TASKS']:
        del background_tasks[id]
    return rv
```

Once we retrieve the resource, it is deleted. This was done by setting `app.config['AUTO_DELETE_BG_TASKS']` to `true` at the top of the app. We will add this decorator to our version endpoints without changing the other part of the code because all of the complexity is hidden in the decorator (how cool is that!):

```
@app.route('/devices/<int:id>/version', methods=['GET'])
@background
def get_device_version(id):
    device = Device.query.get_or_404(id)
    ...
```

```
@app.route('/devices/<device_role>/version', methods=['GET'])
@background
def get_role_version(device_role):
    device_id_list = [device.id for device in Device.query.all() if
device.role == device_role]
...
```

The end result is a two-part process. We will perform the GET request for the endpoint and receive the location header:

```
$ http GET http://172.16.1.173:5000/devices/spine/version
HTTP/1.0 202 ACCEPTED
Content-Length: 2
Content-Type: application/json
Date: <skip>
Location: http://172.16.1.173:5000/status/d02c3f58f4014e96a5dca075e1bb65d4
Server: Werkzeug/0.9.6 Python/3.5.2

{}
```

We can then make a second request to the location to retrieve the result:

```
$ http GET http://172.16.1.173:5000/status/d02c3f58f4014e96a5dca075e1bb65d4
HTTP/1.0 200 OK
Content-Length: 370
Content-Type: application/json
Date: <skip>
Server: Werkzeug/0.9.6 Python/3.5.2

{
  "iosv-1": "('iosv-1', b'show version | i V\r\nCisco IOS Software, IOSv
Software (VIOS-ADVENTERPRISEK9-M), Version 15.6(2)T, RELEASE SOFTWARE
(fc2)\r\n')",
  "iosv-2": "('iosv-2', b'show version | i V\r\nCisco IOS Software, IOSv
Software (VIOS-ADVENTERPRISEK9-M), Version 15.6(2)T, RELEASE SOFTWARE
(fc2)\r\nProcessor board ID 9T7CB2J2V6F0DLWK7V48E\r\n')"
}
```

To verify that the status code 202 is returned when the resource is not ready, we will use the following script, chapter9_request_1.py, to immediately make a request to the new resource:

```
import requests, time

server = 'http://172.16.1.173:5000'
endpoint = '/devices/1/version'

# First request to get the new resource
```

```
r = requests.get(server+endpoint)
resource = r.headers['location']
print("Status: {} Resource: {}".format(r.status_code, resource))

# Second request to get the resource status
r = requests.get(resource)
print("Immediate Status Query to Resource: " + str(r.status_code))

print("Sleep for 2 seconds")
time.sleep(2)
# Third request to get the resource status
r = requests.get(resource)
print("Status after 2 seconds: " + str(r.status_code))
```

As you can see in the result, the status code is returned while the resource is still being run in the background as 202:

```
$ python chapter9_request_1.py
Status: 202 Resource:
http://172.16.1.173:5000/status/1de21f5235c94236a38abd5606680b92
Immediate Status Query to Resource: 202
Sleep for 2 seconds
Status after 2 seconds: 200
```

Our APIs are coming along nicely! Because our network resource is valuable to us, we should secure API access to only authorized personnel. We will add basic security measures to our API in the next section.

Security

For user authentication security, we will use Flask's `httpauth` extension, written by Miguel Grinberg, as well as the password functions in Werkzeug. The `httpauth` extension should have been installed as part of the `requirements.txt` installation at the beginning of this chapter. The new file illustrating the security feature is named `chapter9_9.py`; we will start with a few more module imports:

```
...
from werkzeug.security import generate_password_hash, check_password_hash
from flask.ext.httpauth import HTTPBasicAuth
...
```

We will create an `HTTPBasicAuth` object as well as the `user database` object. Note that, during the user creation process, we will pass the password value; however, we are only storing `password_hash` instead of the `password` itself. This ensures that we are not storing a clear text password for the user:

```python
auth = HTTPBasicAuth()

class User(db.Model):
    __tablename__ = 'users'
    id = db.Column(db.Integer, primary_key=True)
    username = db.Column(db.String(64), index=True)
    password_hash = db.Column(db.String(128))

    def set_password(self, password):
        self.password_hash = generate_password_hash(password)

    def verify_password(self, password):
        return check_password_hash(self.password_hash, password)
```

The `auth` object has a `verify_password` decorator that we can use, along with Flask's `g` global context object that was created when the request started for password verification. Because `g` is global, if we save the user to the `g` variable, it will live through the entire transaction:

```python
@auth.verify_password
def verify_password(username, password):
    g.user = User.query.filter_by(username=username).first()
    if g.user is None:
        return False
    return g.user.verify_password(password)
```

There is a handy `before_request` handler that can be used before any API endpoint is called. We will combine the `auth.login_required` decorator with the `before_request` handler that will be applied to all the API routes:

```python
@app.before_request
@auth.login_required
def before_request():
    pass
```

Lastly, we will use the `unauthorized` error handler to return a `response` object for the 401 unauthorized error:

```python
@auth.error_handler
def unauthorized():
    response = jsonify({'status': 401, 'error': 'unauthorized',
                       'message': 'please authenticate'})
```

```
response.status_code = 401
return response
```

Before we can test user authentication, we will need to create users in our database:

```
>>> from chapter9_9 import db, User
>>> db.create_all()
>>> u = User(username='eric')
>>> u.set_password('secret')
>>> db.session.add(u)
>>> db.session.commit()
>>> exit()
```

Once you start your Flask development server, try to make a request, like we did previously. You should see that, this time, the server will reject the request with a 401 unauthorized error:

```
$ http GET http://172.16.1.173:5000/devices/
HTTP/1.0 401 UNAUTHORIZED
Content-Length: 81
Content-Type: application/json
Date: <skip>
Server: Werkzeug/0.9.6 Python/3.5.2
WWW-Authenticate: Basic realm="Authentication Required"

{
 "error": "unauthorized",
 "message": "please authenticate",
 "status": 401
}
```

We will now need to provide the authentication header for our requests:

```
$ http --auth eric:secret GET http://172.16.1.173:5000/devices/
HTTP/1.0 200 OK
Content-Length: 188
Content-Type: application/json
Date: <skip>
Server: Werkzeug/0.9.6 Python/3.5.2

{
    "device": [
        "http://172.16.1.173:5000/devices/1",
        "http://172.16.1.173:5000/devices/2",
        "http://172.16.1.173:5000/devices/3",
        "http://172.16.1.173:5000/devices/4"
    ]
}
```

We now have a decent RESTful API set up for our network. The user will be able to interact with the APIs now instead of the network devices. They can query for the static content of the network and perform tasks for individual devices or a group of devices. We also added basic security measures to ensure that only the users we created are able to retrieve the information from our API. The cool part is that this is all done within a single file in less than 250 lines of code (less than 200 if you subtract the comments)!

We have now abstracted the underlying vendor API away from our network and replaced them with our own RESTful API. We are free to use what is required in the backend, such as Pexpect, while still providing a uniform frontend to our requester.

Let's take a look at additional resources for Flask so that we can continue to build on our API framework.

Additional resources

Flask is no doubt a feature-rich framework that is growing in features and in the community. We have covered a lot of topics in this chapter, but we have still only scraped the surface of the framework. Besides APIs, you can use Flask for web applications as well as your websites. There are a few improvements that I think we can still make to our network API framework:

- Separate out the database and each endpoint in its own file so that the code is cleaner and easier to troubleshoot.
- Migrate from SQLite to other production-ready databases.
- Use token-based authentication instead of passing the username and password for every transaction. In essence, we will receive a token with finite expiration time upon initial authentication and use the token for further transactions until the expiration.
- Deploy your Flask API app behind a production web server, such as Nginx, along with the Python WSGI server for production use.
- Use an automation process control system, such as Supervisor (http://supervisord.org/), to control the Nginx and Python scripts.

Obviously, the recommended improvement choices will vary greatly from company to company. For example, the choice of database and web server may have implications for the company's technical preference as well as the other teams' input. The use of token-based authentication might not be necessary if the API is only used internally and other forms of security have been put into place. For these reasons, I would like to provide you with additional links as extra resources should you choose to move forward with any of the preceding items.

Here are some of the links I find useful when thinking about design patterns, database options, and general Flask features:

- Best practices on Flask design patterns: `http://flask.pocoo.org/docs/0.10/patterns/`
- Flask API: `http://flask.pocoo.org/docs/0.12/api/`
- Deployment options: `http://flask.pocoo.org/docs/0.12/deploying/`

Due to the nature of Flask and the fact that it relies on the extension outside of its small core, sometimes, you might find yourself jumping from one document to another. This can be frustrating, but the upside is that you only need to know about the extension you are using, which I feel saves time in the long run.

Summary

In this chapter, we started to move onto the path of building REST APIs for our network. We looked at different popular Python web frameworks, namely Django and Flask, and compared and contrasted the two. By choosing Flask, we are able to start small and expand on features by using Flask extensions.

In our lab, we used the virtual environment to separate the Flask installation base from our global site-packages. The lab network consists of four nodes, two of which we have designated as spine routers while the other two are designated as leaf routers. We took a tour of the basics of Flask and used the simple HTTPie client for testing our API setup.

Among the different setups of Flask, we placed special emphasis on URL dispatch as well as the URL variables because they are the initial logic between the requesters and our API system. We took a look at using Flask-SQLAlchemy and SQLite to store and return network elements that are static in nature. For operation tasks, we also created API endpoints while calling other programs, such as Pexpect, to accomplish configuration tasks. We improved the setup by adding asynchronous handling as well as user authentication to our API. Toward the end of this chapter, we looked at some of the additional resource links we can follow to add even more security and other features.

In `Chapter 10`, *AWS Cloud Networking*, we will shift our gear to look at cloud networking using **Amazon Web Services (AWS)**.

10 AWS Cloud Networking

Cloud computing is one of the major trends in computing today. Public cloud providers have transformed the high-tech industry and what it means to launch a service from scratch. We no longer need to build our own infrastructure; we can pay the public cloud providers to rent a portion of their resources for our infrastructure needs. Nowadays, walking around any technology conferences or meetups, we will be hard-pressed to find a person who has not learned about, used, or built services based in the cloud. Cloud computing is here, and we better get used to working with it.

There are several service models of cloud computing, roughly divided into **Software-as-a-Service (SaaS)** (https://en.wikipedia.org/wiki/Software_as_a_service), **Platform-as-a-Service (PaaS)** (https://en.wikipedia.org/wiki/Cloud_computing#Platform_as_a_service_(PaaS)), and **Infrastructure-as-a-Service (IaaS)** (https://en.wikipedia.org/wiki/Infrastructure_as_a_service). Each service model offers a different level of abstraction from the user's perspective. For us, networking is part of the Infrastructure-as-a-Service offering and the focus of this chapter.

Amazon Web Services (AWS—https://aws.amazon.com/) is the first company to offer IaaS public cloud services and the clear leader in the space by market share in 2018. If we define the term **Software Defined Networking (SDN)** as a group of software services working together to create network constructs – IP addresses, access lists, Network Address Translation, routers – we can make the argument that AWS is the world's largest implementation of SDN. They utilize their massive scale of the global network, data centers, and hosts to offer an amazing array of networking services.

AWS Cloud Networking

 If you are interested in learning about Amazon's scale and networking, I would highly recommend taking a look at James Hamilton's AWS re:Invent 2014 talk: https://www.youtube.com/watch?v=JIQETrFC_SQ. It is a rare insider's view of the scale and innovation at AWS.

In this chapter, we will discuss the networking services offered by the AWS cloud services and how we can use Python to work with them:

- AWS setup and networking overview
- Virtual private cloud
- Direct Connect and VPN
- Networking scaling services
- Other AWS network services

AWS setup

If you do not already have an AWS account and wish to follow along with these examples, please log on to https://aws.amazon.com/ and sign up. The process is pretty straightforward and simple; you will need a credit card and some form of verification. AWS offers a number of services in a free tier (https://aws.amazon.com/free/), where you can use some of the most popular services for free up to a certain level.

Some of the services listed are free for the first year, and others are free up to a certain limit without time restraint. Please check the AWS site for the latest offerings:

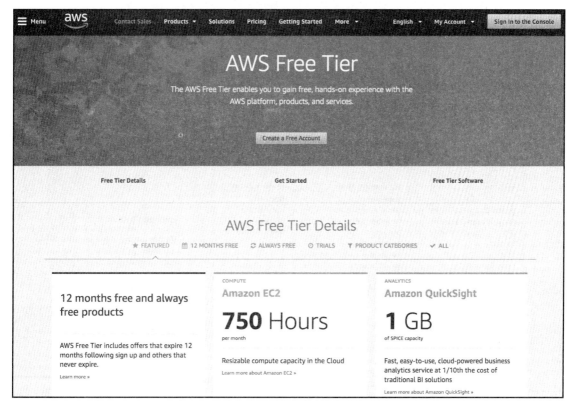

AWS free tier

Once you have an account, you can sign in via the AWS console (`https://console.aws.amazon.com/`) and take a look at the different services offered by AWS. The console is where we can configure all the services and look at our monthly bills:

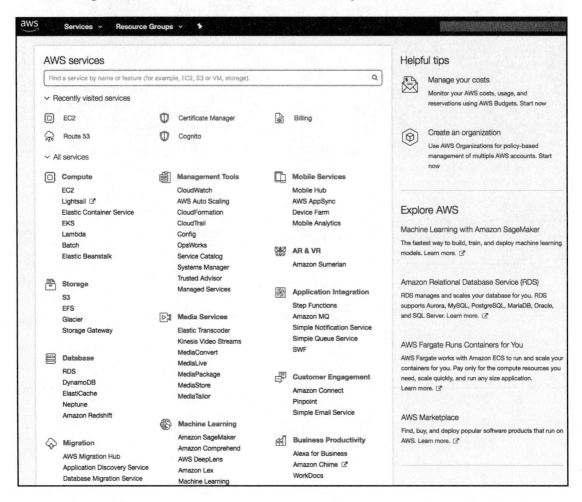

AWS console

AWS CLI and Python SDK

We can also manage the AWS services via the command-line interface. The AWS CLI is a Python package that can be installed via PIP (https://docs.aws.amazon.com/cli/latest/userguide/installing.html). Let's install it on our Ubuntu host:

```
$ sudo pip3 install awscli
$ aws --version
aws-cli/1.15.59 Python/3.5.2 Linux/4.15.0-30-generic botocore/1.10.58
```

Once the AWS CLI is installed, for easier and more secure access, we will create a user and configure AWS CLI with the user credentials. Let's go back to the AWS console and select **IAM** for user and access management:

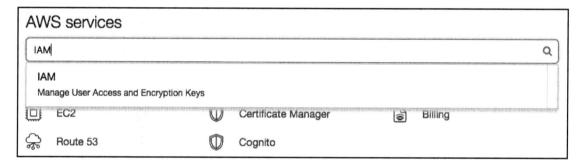

AWS IAM

AWS Cloud Networking

We can choose `Users` on the left panel to create a user:

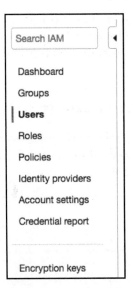

Select programmatic access and assign the user to the default administrator group:

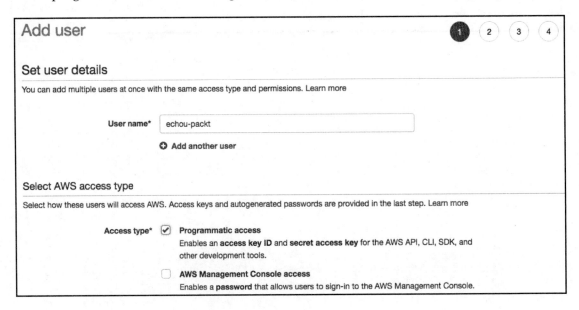

The last step will show an **Access key ID** and a **Secret access key**. Copy them into a text file and keep it in a safe place:

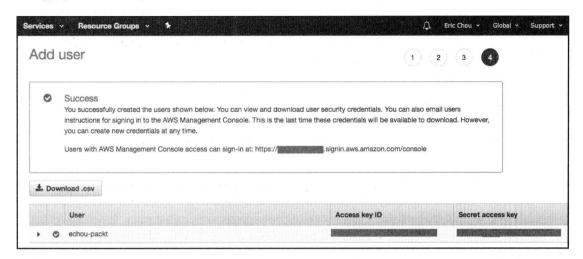

We will complete the AWS CLI authentication credential setup via `aws configure` in the terminal. We will go over AWS regions in the upcoming section; we will use `us-east-1` for now, but feel free to come back and change this value later:

```
$ aws configure
AWS Access Key ID [None]: <key>
AWS Secret Access Key [None]: <secret>
Default region name [None]: us-east-1
Default output format [None]: json
```

We will also install the AWS Python SDK, Boto3 (https://boto3.readthedocs.io/en/latest/):

```
$ sudo pip install boto3
$ sudo pip3 install boto3

# verification
$ python3
Python 3.5.2 (default, Nov 23 2017, 16:37:01)
[GCC 5.4.0 20160609] on linux
Type "help", "copyright", "credits" or "license" for more information.
>>> import boto3
>>> exit()
```

We are now ready to move on to the subsequent sections, starting with an introduction to AWS cloud networking services.

AWS network overview

When we discuss AWS services, we need to start at the top with regions and availability zones. They have big implications for all of our services. At the time of writing this book, AWS listed 18 Regions, 55 **Availability Zones (AZ)**, and one local region around the world. In the words of AWS Global Infrastructure, (`https://aws.amazon.com/about-aws/global-infrastructure/`):

> "The AWS Cloud infrastructure is built around Regions and Availability Zones (AZs). AWS Regions provide multiple, physically separated and isolated Availability Zones which are connected with low latency, high throughput, and highly redundant networking."

Some of the services AWS offer are global, but most of the services are region-based. What this means for us is that we should build our infrastructure in a region that is closest to our intended users. This will reduce the latency of the service to our customer. If our users are in the United States east coast, we should pick `us-east-1` (N. Virginia) or `us-east-2` (Ohio) as our region if the service is regional-based:

#	Region & Number of Availability Zones		○	New Region (coming soon)
	US East N. Virginia (6), Ohio (3)	**China** Beijing (2), Ningxia (3)		Bahrain Hong Kong SAR, China Sweden AWS GovCloud (US-East)
	US West N. California (3), Oregon (3)	**Europe** Frankfurt (3), Ireland (3), London (3), Paris (3)		
	Asia Pacific Mumbai (2), Seoul (2), Singapore (3), Sydney (3), Tokyo (4), Osaka-Local (1)[1]	**South America** São Paulo (3) **AWS GovCloud (US-West) (3)**		
	Canada Central (2)			

AWS regions

Not all regions are available to all users, for example, GovCloud and the **China** region are not available to users in the United States by default. You can list the regions available to you via `aws ec2 describe-regions`:

```
$ aws ec2 describe-regions
{
    "Regions": [
        {
            "RegionName": "ap-south-1",
            "Endpoint": "ec2.ap-south-1.amazonaws.com"
        },
        {
            "RegionName": "eu-west-3",
            "Endpoint": "ec2.eu-west-3.amazonaws.com"
        },
...
```

All the regions are completely independent of one another. Most resources are not replicated across regions. If we have multiple regions, say `US-East` and `US-West`, and need redundancy between them, we will need to replicate the necessary resources ourselves. The way you choose a region is on the top right corner of the console:

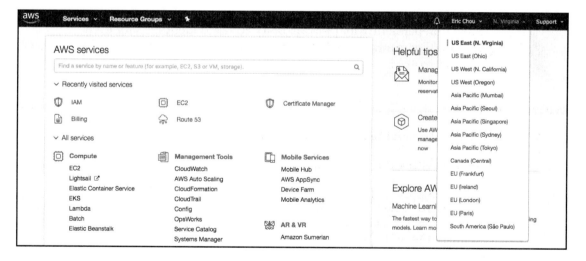

 If the service is region-based, for example, EC2, the portal will only show the service when the right region is selected. If our EC2 instances are in `us-east-1` and we are looking at the us-west-1 portal, none of the EC2 instances will show up. I have made this mistake a few times, and wondered where all of my instances went!

AWS Cloud Networking

The number behind the regions in the preceding AWS regions screenshot represents the number of AZ in each region. Each region has multiple availability zones. Each availability zone is isolated, but the AZs in a region are connected through low-latency fiber connections:

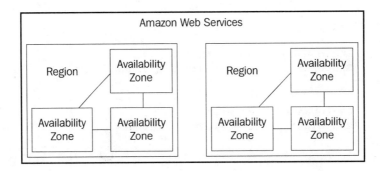

AWS regions and availability zones

Many of the resources we built are copied across availability zones. The concept of AZ is very important, and its constraints are important to us for the network services we will build.

 AWS independently maps availability zones to identifiers for each account. For example, my availability zone, us-eas-1a, might not be the same as `us-east-1a` for another account.

We can check the AZs in a region in AWS CLI:

```
$ aws ec2 describe-availability-zones --region us-east-1
{
    "AvailabilityZones": [
        {
            "Messages": [],
            "RegionName": "us-east-1",
            "State": "available",
            "ZoneName": "us-east-1a"
        },
        {
            "Messages": [],
            "RegionName": "us-east-1",
            "State": "available",
            "ZoneName": "us-east-1b"
        },
...
```

[326]

Why do we care about regions and availability zones so much? As we will see in the coming few sections, the networking services are usually bound by the region and availability zones. **Virtual Private Cloud (VPC)**, for example, needs to reside entirely in one region, and each subnet needs to reside entirely in one availability zone. On the other hand, **NAT Gateway** is AZ-bound, so we will need to create one per AZ if we needed redundancy. We will go over both services in more detail, but their use cases are offered here as examples of how regions and availability zones are the basis of the AWS network services offering.

AWS Edge locations are part of the **AWS CloudFront** content delivery network in 59 cities across 26 countries. These edge locations are used to distribute content with low latency with a smaller footprint than the full data center Amazon builds for the region and availability zones. Sometimes, people mistook the edge locations' point-of-presence for full AWS regions. If the footprint is listed as an edge location only, the AWS services such as EC2 or S3 will not be offered. We will revisit the edge location in the *AWS CloudFront* section.

AWS Transit Centers is one of the least documented aspects of AWS networks. It was mentioned in James Hamilton's 2014 **AWS re:Invent** keynote (`https://www.youtube.com/watch?v=JIQETrFC_SQ`) as the aggregation points for different AZs in the region. To be fair, we do not know if the transit center still exists and functions the same way after all these years. However, it is fair to make an educated guess about the placement of the transit center and its correlation about the **AWS Direct Connect** service that we will look at later in this chapter.

James Hamilton, a VP and distinguished engineer from AWS, is one of the most influential technologists at AWS. If there is anybody who I would consider authoritative when it comes to AWS networking, it would be him. You can read more about his visions on his blog, Perspectives, at `https://perspectives.mvdirona.com/`.

It is impossible to cover all of the services related to AWS in one chapter. There are some relevant services not directly related to networking that we do not have the space to cover, but we should be familiar with:

- The **Identify and Access Management (IAM)** service, `https://aws.amazon.com/iam/`, is the service that enables us to manage access to AWS services and resources securely.

AWS Cloud Networking

- **Amazon Resource Names (ARNs)**, https://docs.aws.amazon.com/general/latest/gr/aws-arns-and-namespaces.html, uniquely identify AWS resources across all of AWS. This resource name is important when we need to identify a service, such as DynamoDB and API Gateway, that needs access to our VPC resources.
- **Amazon Elastic Compute Cloud (EC2)**, https://aws.amazon.com/ec2/, is the service that enables us to obtain and provision compute capacities, such as Linux and Windows instances, via AWS interfaces. We will use EC2 instances throughout this chapter in our examples.

For the sake of learning, we will exclude AWS GovCloud (US) and China, neither of which uses the AWS global infrastructure and have their own limitations.

This was a relatively long introduction to the AWS network overview, but an important one. These concepts and terms will be referred to in the rest of the chapters in this book. In the upcoming section, we will take a look at the most import concept (in my opinion) for AWS networking: the virtual private cloud.

Virtual private cloud

Amazon Virtual Private Cloud (Amazon VPC) enables customers to launch AWS resources into a virtual network dedicated to the customer's account. It is truly a customizable network that allows you to define your own IP address range, add and delete subnets, create routes, add VPN gateways, associate security policies, connect EC2 instances to your own datacenter, and much more. In the early days when VPC was not available, all EC2 instances in the AZ were on a single, flat network that was shared among all customers. How comfortable would the customer be with putting their information in the cloud? Not very, I'd imagine. Between the launch of EC2 in 2007 until the launch of VPC in 2009, VPC functions was one of the most requested features of AWS.

The packets leaving your EC2 host in a VPC are intercepted by the Hypervisor. The Hypervisor will check them with a mapping service which understands our VPC construct. The packets leaving your EC2 hosts are encapsulated with the AWS real servers' source and destination addresses. The encapsulation and mapping service allows for the flexibility of VPC, but also some of the limitations (multicast, sniffing) of VPC. This is, after all, a virtual network.

Since December 2013, all EC2 instances are VPC-only. If we use a launch wizard to create our EC2 instance, it will automatically be put into a default VPC with a virtual internet gateway for public access. In my opinion, all but the most basic use cases should use the default VPC. For most cases, we would need to define our non-default, customized VPC.

Let's create the following VPC using the AWS console in `us-east-1`:

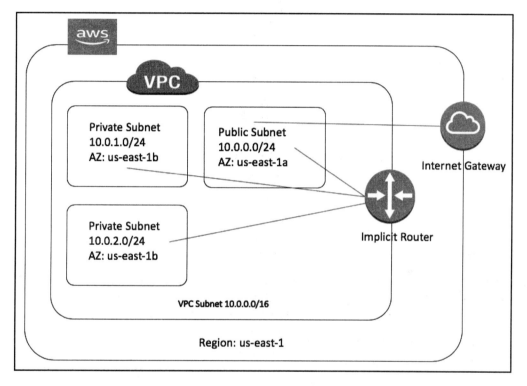

Our first VPC in US-East-1

If you recall, VPC is AWS region-bound, and the subnets are Availability Zone-based. Our first VPC will be based in `us-east-1`; the three subnets will be allocated to three different availability zones in 1a, 1b, and 1c.

Using the AWS console to create the VPC and subnets is pretty straightforward, and AWS provides a number of good tutorials online. I have listed the steps with the associated links on the VPC dashboard:

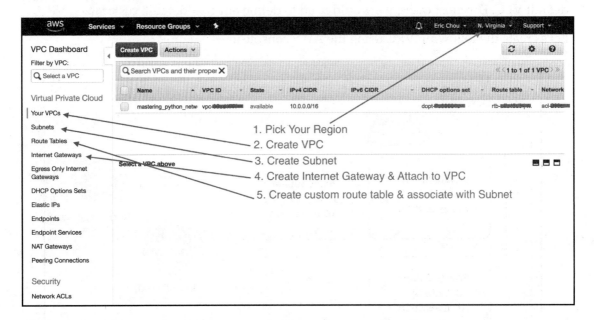

The first two steps are point and click processes that most network engineers can work through, even without prior experience. By default, the VPC only contains the local route, `10.0.0.0/16`. Now, we will create an internet gateway and associate it with the VPC:

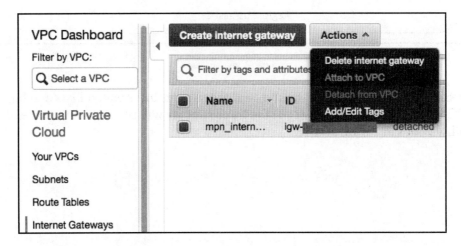

Chapter 10

We can then create a custom route table with a default route pointing to the internet gateway. We will associate this route table with our subnet in us-east-1a, 10.0.0.0/24, thus allowing it to be public facing:

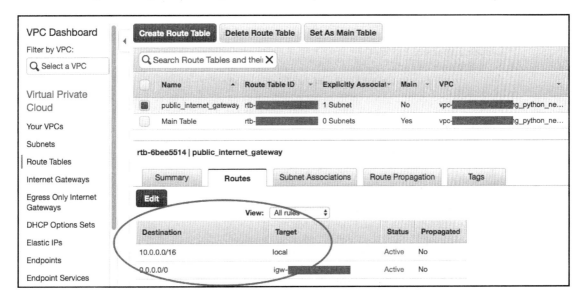

Route table

Let's use Boto3 Python SDK to see what we have created; I used the tag mastering_python_networking_demo as the tag for the VPC, which we can use as the filer:

```
$ cat Chapter10_1_query_vpc.py
#!/usr/bin/env python3

import json, boto3

region = 'us-east-1'
vpc_name = 'mastering_python_networking_demo'

ec2 = boto3.resource('ec2', region_name=region)
client = boto3.client('ec2')

filters = [{'Name':'tag:Name', 'Values':[vpc_name]}]

vpcs = list(ec2.vpcs.filter(Filters=filters))
for vpc in vpcs:
    response = client.describe_vpcs(
                VpcIds=[vpc.id,]
```

AWS Cloud Networking

```
            )
    print(json.dumps(response, sort_keys=True, indent=4))
```

This script will allow us to programmatically query the region for the VPC we created:

```
$ python3 Chapter10_1_query_vpc.py
{
    "ResponseMetadata": {
        "HTTPHeaders": {
            "content-type": "text/xml;charset=UTF-8",
            ...
        },
        "HTTPStatusCode": 200,
        "RequestId": "48e19be5-01c1-469b-b6ff-9c45f2745483",
        "RetryAttempts": 0
    },
    "Vpcs": [
        {
            "CidrBlock": "10.0.0.0/16",
            "CidrBlockAssociationSet": [
                {
                    "AssociationId": "...",
                    "CidrBlock": "10.0.0.0/16",
                    "CidrBlockState": {
                        "State": "associated"
                    }
                }
            ],
            "DhcpOptionsId": "dopt-....",
            "InstanceTenancy": "default",
            "IsDefault": false,
            "State": "available",
            "Tags": [
                {
                    "Key": "Name",
                    "Value": "mastering_python_networking_demo"
                }
            ],
            "VpcId": "vpc-...."
        }
    ]
}
```

 The Boto3 VPC API documentation can be found at https://boto3.readthedocs.io/en/latest/reference/services/ec2.html#vpc.

[332]

You may be wondering about how the subnets can reach one another within the VPC. In a physical network, the network needs to connect to a router to reach beyond its own local network. It is not so different in VPC, except it is an *implicit router* with a default routing table of the local network, which in our example is 10.0.0.0/16. This implicit router was created when we created our VPC.

Route tables and route targets

Routing is one of the most important topics in network engineering. It is worth looking at it more closely. We already saw that we had an implicit router and the main routing table when we created the VPC. From the last example, we created an internet gateway, a custom routing table with a default route pointing to the internet gateway, and associated the custom routing table with a subnet.

The concept of the route target is where VPC is a bit different than traditional networking. In summary:

- Each VPC has an implicit router
- Each VPC has the main routing table with the local route populated
- You can create custom-routing tables
- Each subnet can follow a custom-routing table or the default main routing table
- The route table route target can be an internet gateway, NAT gateway, VPC peers, and so on

We can use Boto3 to look at the custom route tables and association with the subnets:

```
$ cat Chapter10_2_query_route_tables.py
#!/usr/bin/env python3

import json, boto3

region = 'us-east-1'
vpc_name = 'mastering_python_networking_demo'

ec2 = boto3.resource('ec2', region_name=region)
client = boto3.client('ec2')

response = client.describe_route_tables()
print(json.dumps(response['RouteTables'][0], sort_keys=True, indent=4))
```

We only have one custom route table:

```
$ python3 Chapter10_2_query_route_tables.py
{
    "Associations": [
        {
            ....
        }
    ],
    "PropagatingVgws": [],
    "RouteTableId": "rtb-6bee5514",
    "Routes": [
        {
            "DestinationCidrBlock": "10.0.0.0/16",
            "GatewayId": "local",
            "Origin": "CreateRouteTable",
            "State": "active"
        },
        {
            "DestinationCidrBlock": "0.0.0.0/0",
            "GatewayId": "igw-...",
            "Origin": "CreateRoute",
            "State": "active"
        }
    ],
    "Tags": [
        {
            "Key": "Name",
            "Value": "public_internet_gateway"
        }
    ],
    "VpcId": "vpc-..."
}
```

Creating the subnets are straight forward by clicking on the left subnet section and follow the on-screen instruction. For our purpose, we will create three subnets, `10.0.0.0/24` public subnet, `10.0.1.0/24`, and `10.0.2.0/24` private subnets.

We now have a working VPC with three subnets: one public and two private. So far, we have used the AWS CLI and Boto3 library to interact with AWS VPC. Let's take a look at another automation tool, **CloudFormation**.

Automation with CloudFormation

AWS CloudFomation (`https://aws.amazon.com/cloudformation/`), is one way in which we can use a text file to describe and launch the resource that we need. We can use CloudFormation to provision another VPC in the `us-west-1` region:

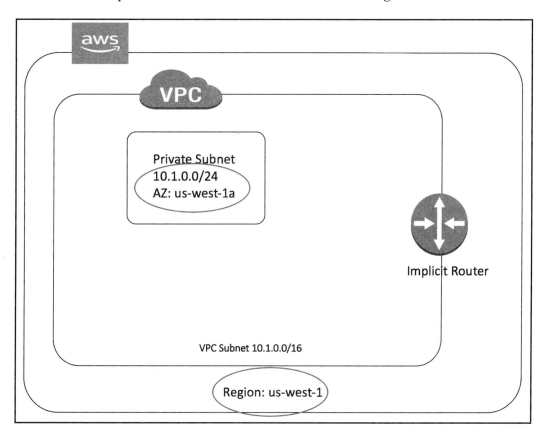

VPC for US-West-1

The CloudFormation template can be in YAML or JSON; we will use YAML for our first template for provisioning:

```
$ cat Chapter10_3_cloud_formation.yml
AWSTemplateFormatVersion: '2010-09-09'
Description: Create VPC in us-west-1
Resources:
  myVPC:
    Type: AWS::EC2::VPC
    Properties:
```

```
        CidrBlock: '10.1.0.0/16'
        EnableDnsSupport: 'false'
        EnableDnsHostnames: 'false'
        Tags:
          - Key: Name
            Value: 'mastering_python_networking_demo_2'
```

We can execute the template via the AWS CLI. Notice that we specify a region of us-west-1 in our execution:

```
$ aws --region us-west-1 cloudformation create-stack --stack-name 'mpn-ch10-demo' --template-body file://Chapter10_3_cloud_formation.yml
{
    "StackId": "arn:aws:cloudformation:us-west-1:<skip>:stack/mpn-ch10-demo/<skip>"
}
```

We can verify the status via AWS CLI:

```
$ aws --region us-west-1 cloudformation describe-stacks --stack-name mpn-ch10-demo
{
    "Stacks": [
        {
            "CreationTime": "2018-07-18T18:45:25.690Z",
            "Description": "Create VPC in us-west-1",
            "DisableRollback": false,
            "StackName": "mpn-ch10-demo",
            "RollbackConfiguration": {},
            "StackStatus": "CREATE_COMPLETE",
            "NotificationARNs": [],
            "Tags": [],
            "EnableTerminationProtection": false,
            "StackId": "arn:aws:cloudformation:us-west-1<skip>"
        }
    ]
}
```

For demonstration purposes, the last CloudFormation template created a VPC without any subnet. Let's delete that VPC and use the following template to create both the VPC as well as the subnet. Notice that we will not have the VPC-id before VPC creation, so we will use a special variable to reference the VPC-id in the subnet creation. This is the same technique we can use for other resources, such as the routing table and internet gateway:

```
$ cat Chapter10_4_cloud_formation_full.yml
AWSTemplateFormatVersion: '2010-09-09'
Description: Create subnet in us-west-1
Resources:
```

```
  myVPC:
    Type: AWS::EC2::VPC
    Properties:
      CidrBlock: '10.1.0.0/16'
      EnableDnsSupport: 'false'
      EnableDnsHostnames: 'false'
      Tags:
        - Key: Name
          Value: 'mastering_python_networking_demo_2'
  mySubnet:
    Type: AWS::EC2::Subnet
    Properties:
      VpcId: !Ref myVPC
      CidrBlock: '10.1.0.0/24'
      AvailabilityZone: 'us-west-1a'
      Tags:
        - Key: Name
          Value: 'mpn_demo_subnet_1'
```

We can execute and verify the creation of the resources as follows:

```
$ aws --region us-west-1 cloudformation create-stack --stack-name mpn-ch10-demo-2 --template-body file://Chapter10_4_cloud_formation_full.yml
{
    "StackId": "arn:aws:cloudformation:us-west-1:<skip>:stack/mpn-ch10-demo-2/<skip>"
}

$ aws --region us-west-1 cloudformation describe-stacks --stack-name mpn-ch10-demo-2
{
    "Stacks": [
        {
            "StackStatus": "CREATE_COMPLETE",
            ...
            "StackName": "mpn-ch10-demo-2",
            "DisableRollback": false
        }
    ]
}
```

AWS Cloud Networking

We can also verify the VPC and subnet information from the AWS console. We will verify the VPC from the console first:

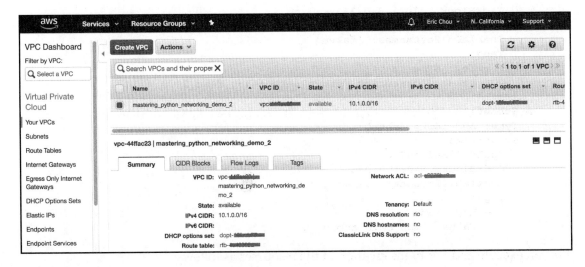

VPC in us-west-1

We can also take a look at the subnet:

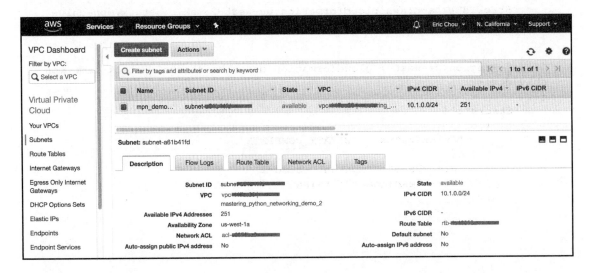

Subnet in us-west-1

Chapter 10

We now have two VPCs in the two coasts of the United States. They are currently behaving like two islands, each by themselves. This may or may not be your desired state of operation. If you would like the to VPC to be able to connect them to each other, we can use VPC peering (https://docs.aws.amazon.com/AmazonVPC/latest/PeeringGuide/vpc-peering-basics.html) to allow direct communication.

VPC peering is not limited to the same account. You can connect VPCs across different accounts, as long as the request was accepted and the other aspects (security, routing, DNS name) are taken care of.

In the coming section, we will take a look at VPC security groups and the network access control list.

Security groups and the network ACL

AWS **Security Groups** and the **Access Control** list can be found under the **Security** section of your VPC:

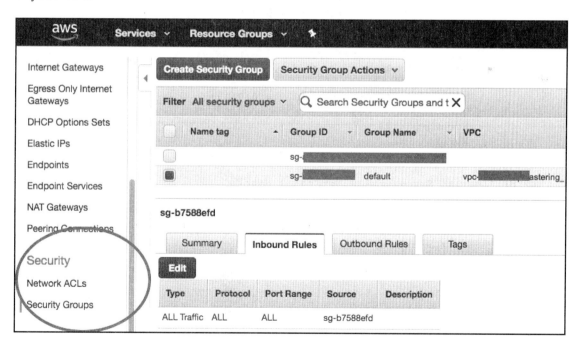

VPC security

[339]

A security group is a stateful virtual firewall that controls inbound and outbound access for resources. Most of the time, we will use the security group as a way to limit public access to our EC2 instance. The current limitation is 500 security groups in each VPC. Each security group can contain up to 50 inbound and 50 outbound rules. You can use the following sample script to create a security group and two simple ingress rules:

```python
$ cat Chapter10_5_security_group.py
#!/usr/bin/env python3

import boto3

ec2 = boto3.client('ec2')

response = ec2.describe_vpcs()
vpc_id = response.get('Vpcs', [{}])[0].get('VpcId', '')

# Query for security group id
response = ec2.create_security_group(GroupName='mpn_security_group',
                                      Description='mpn_demo_sg',
                                      VpcId=vpc_id)
security_group_id = response['GroupId']
data = ec2.authorize_security_group_ingress(
    GroupId=security_group_id,
    IpPermissions=[
        {'IpProtocol': 'tcp',
         'FromPort': 80,
         'ToPort': 80,
         'IpRanges': [{'CidrIp': '0.0.0.0/0'}]},
        {'IpProtocol': 'tcp',
         'FromPort': 22,
         'ToPort': 22,
         'IpRanges': [{'CidrIp': '0.0.0.0/0'}]}
    ])
print('Ingress Successfully Set %s' % data)

# Describe security group
#response = ec2.describe_security_groups(GroupIds=[security_group_id])
print(security_group_id)
```

We can execute the script and receive confirmation on the creation of the security group that can be associated with other AWS resources:

```
$ python3 Chapter10_5_security_group.py
Ingress Successfully Set {'ResponseMetadata': {'RequestId': '<skip>',
'HTTPStatusCode': 200, 'HTTPHeaders': {'server': 'AmazonEC2', 'content-
type': 'text/xml;charset=UTF-8', 'date': 'Wed, 18 Jul 2018 20:51:55 GMT',
'content-length': '259'}, 'RetryAttempts': 0}}
sg-<skip>
```

Network **Access Control Lists (ACLs)** is an additional layer of security that is stateless. Each subnet in VPC is associated with a network ACL. Since ACL is stateless, you will need to specify both inbound and outbound rules.

The important differences between the security group and ACLs are as follows:

- The security group operates at the network interface level where ACL operates at the subnet level
- For a security group, we can only specify allow rules but not deny rules, whereas ACL supports both allow and deny rules
- A security group is stateful; return traffic is automatically allowed. Return traffic needs to be specifically allowed in ACL

Let's take a look at one of the coolest feature of AWS networking, Elastic IP. When I initially learned about Elastic IP, I was blown away by the ability of assigning and reassigning IP addresses dynamically.

Elastic IP

Elastic IP (EIP) is a way to use a public IPv4 address that's reachable from the internet. It can be dynamically assigned to an EC2 instance, network interface, or other resources. A few characteristics of Elastic IP are as follows:

- The Elastic IP is associated with the account and is region-specific. For example, the EIP in `us-east-1` can only be associated with resources in `us-east-1`.
- You can disassociate an Elastic IP from a resource, and re-associate it with a different resource. This flexibility can sometimes be used to ensure high availability. For example, you can migrate from a smaller EC2 instance to a larger EC2 instance by reassigning the same IP address from the small EC2 instance to the larger one.
- There is a small hourly charge associated with Elastic IP.

You can request **Elastic IP** from the portal. After assignment, you can associate it with the desired resources:

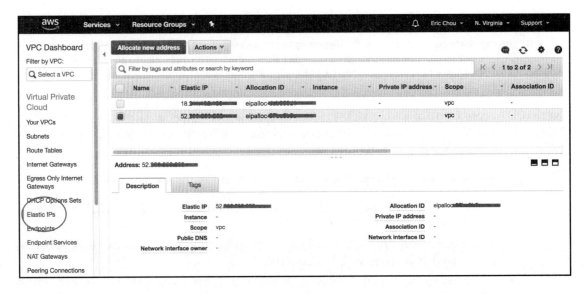

Elastic IP

 Unfortunately, Elastic IP has a default limit of five per region, `https://docs.aws.amazon.com/vpc/latest/userguide/amazon-vpc-limits.html`.

In the coming section, we will look at how we can use the NAT Gateway to allow communication for the private subnets to the internet.

NAT Gateway

To allow the hosts in our EC2 public subnet to be accessed from the internet, we can allocate an Elastic IP and associate it with the network interface of the EC2 host. However, at the time of writing this book, there is a limit of five Elastic IPs per EC2-VPC (`https://docs.aws.amazon.com/AmazonVPC/latest/UserGuide/VPC_Appendix_Limits.html#vpc-limits-eips`). Sometimes, it would be nice to allow the host in a private subnet outbound access when needed without creating a permanent one-to-one mapping between the Elastic IP and the EC2 host.

Chapter 10

This is where **NAT Gateway** can help, by allowing the hosts in the private subnet temporarily outbound access by performing a **Network Address Translation (NAT)**. This operation is similar to the **Port Address Translation (PAT)** that we normally perform on the corporate firewall. To use a NAT Gateway, we can perform the following steps:

- Create a NAT Gateway in a subnet with access to the internet gateway via the AWS CLI, Boto3 library, or AWS console. The NAT Gateway will need to be assigned with an Elastic IP.
- Point the default route in the private subnet to the NAT Gateway.
- The NAT Gateway will follow the default route to the internet gateway for external access.

This operation can be illustrated in the following diagram:

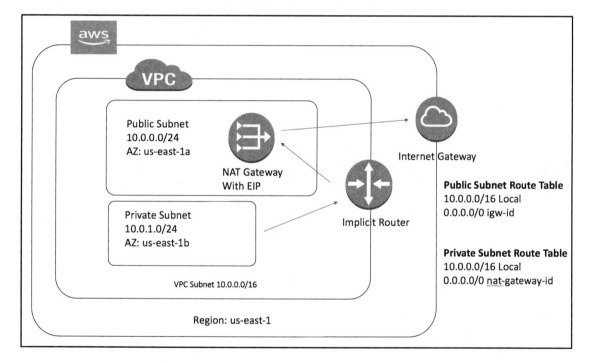

NAT Gateway operations

AWS Cloud Networking

One of the most common questions for NAT Gateway typically surrounds which subnet the NAT Gateway should reside in. The rule of thumb is to remember that the NAT Gateway needs public access. Therefore, it should be created in the subnet with public internet access with an available Elastic IP to be assigned to it:

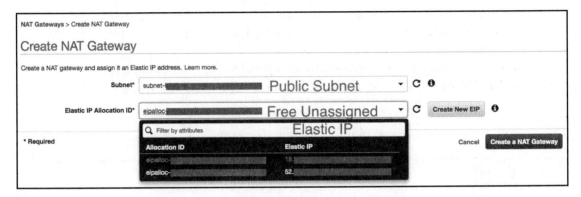

NAT Gateway creation

In the coming section, we will take a look at how to connect our shiny virtual network in AWS to our physical network.

Direct Connect and VPN

Up to this point, our VPC is a self-contained network that resides in the AWS network. It is flexible and functional, but to access the resources inside of the VPC, we will need to access them with their internet-facing services such as SSH and HTTPS.

In this section, we will look at the two ways AWS allow us to connect to the VPC from our private network: IPSec VPN Gateway and Direct Connect.

VPN Gateway

The first way to connect our on-premise network to VPC is with traditional IPSec VPN connections. We will need a publicly accessible device that can establish VPN connections to AWS's VPN device. The customer gateway needs to support route-based IPSec VPNs where the VPN connection is treated as a connection that a routing protocol can run over the virtual link. Currently, AWS recommends using BGP to exchange routes.

On the VPC side, we can follow a similar routing table where we can route a particular subnet toward the **Virtual Private Gateway** target:

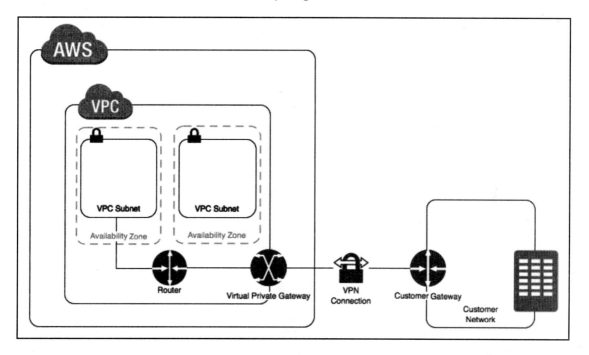

VPC VPN connection (source: https://docs.aws.amazon.com/AmazonVPC/latest/UserGuide/VPC_VPN.html)

Besides IPSec VPN, we can also use a dedicated circuit to connect.

Direct Connect

The IPSec VPN connection we saw is an easy way to provide connectivity for on-premise equipment to the AWS cloud resources. However, it suffers the same faults that IPSec over the internet always does: it is unreliable, and we have very little control over it. There is very little performance monitoring and no **Service-Level Agreement** (**SLA**) until the connection reaches a part of the internet that we can control.

For all of these reasons, any production-level, mission-critical traffic is more likely to traverse through the second option Amazon provides, that is, AWS Direct Connect. AWS Direct Connect allows customers to connect their data center and colocation to their AWS VPC with a dedicated virtual circuit. The somewhat difficult part of this operation is usually bringing our network to where we can connect with AWS physically, typically in a carrier hotel. You can find a list of the AWS Direct Connect locations here: https://aws.amazon.com/directconnect/details/. The Direct Connect link is just a fiber patch connection that you can order from the particular carrier hotel to patch the network to a network port and configure the dot1q trunk's connectivity.

There are also increasingly more connectivity options for Direct Connect via a third-party carrier with MPLS circuits and aggregated links. One of the most affordable options that I found and use is the Equinix Cloud Exchange (https://www.equinix.com/services/interconnection-connectivity/cloud-exchange/). By using the Equinix Cloud Exchange, we can leverage the same circuit and connect to different cloud providers at a fraction of the cost of dedicated circuits:

Equinix Cloud Exchange (source: https://www.equinix.com/services/interconnection-connectivity/cloud-exchange/)

In the upcoming section, we will take a look at some of the network scaling services AWS offers.

Network scaling services

In this section, we will take a look at some of the network services AWS offers. Many of the services do not have a direct network implication, such as DNS and content distribution network. They are relevant in our discussion due to their close relationship with the network and application's performance.

Elastic Load Balancing

Elastic Load Balancing (ELB) allows incoming traffic from the internet to be automatically distributed across multiple EC2 instances. Just like load balancers in the physical world, this allows us to have better redundancy and fault tolerance while reducing the per-server load. ELB comes in two flavors: application and network load balancing.

The application load balancer handles web traffic via HTTP and HTTPS; the network load balancer operates on a TCP level. If your application runs on HTTP or HTTPS, it is generally a good idea to go with the application load balancer. Otherwise, using the network load balancer is a good bet.

A detailed comparison of the application and network load balancer can be found at https://aws.amazon.com/elasticloadbalancing/details/:

Comparison of Elastic Load Balancing Products

You can select the appropriate load balancer based on your application needs. If you need flexible application management, we recommend that you use an **Application Load Balancer**. If extreme performance and static IP is needed for your application, we recommend that you use a **Network Load Balancer**. If you have an existing application that was built within the EC2-Classic network, then you should use a **Classic Load Balancer**.

Feature	Application Load Balancer	Network Load Balancer	Classic Load Balancer
Protocols	HTTP, HTTPS	TCP	TCP, SSL, HTTP, HTTPS
Platforms	VPC	VPC	EC2-Classic, VPC
Health checks	✔	✔	✔
CloudWatch metrics	✔	✔	✔
Logging	✔	✔	✔
Zonal fail-over	✔	✔	✔

Elastic Load Balancer Comparison (Source: https://aws.amazon.com/elasticloadbalancing/details/)

Elastic Load Balancer offers a way to load balance traffic once it enters the resource in our region. The AWS Route53 DNS service allows geographic load balance between regions.

Route53 DNS service

We all know what domain name services are; Route53 is AWS's DNS service. Route53 is a full-service domain registrar where you can purchase and manage domains directly from AWS. Regarding network services, DNS allows a way to load balance between geographic regions by service domain names in a round-robin fashion between regions.

We need the following items before we can use DNS for load balancing:

- An Elastic Load Balancer in each of the intended load balance regions.
- A registered domain name. We do not need Route53 as the domain registrar.
- Route53 is the DNS service for the domain.

We can then use the Route 53 latency-based routing policy with health-check in an active-active environment between the two Elastic Load Balancers.

CloudFront CDN services

CloudFront is Amazon's **Content Delivery Network** (**CDN**) that reduces the latency for content delivery by physically serving the content closer to the customer. The content can be static web page content, videos, applications, APIs, or most recently, Lambda functions. CloudFront edge locations include the existing AWS regions, but are also in many other locations around the globe. The high-level operation of CloudFront is as follows:

- Users access your website for one or more objects
- DNS routes the request to the Amazon CloudFront edge location closest to the user's request
- The CloudFront edge location will either service the content via the cache or request the object from the origin

AWS CloudFront and CDN services in general are typically handled by application developers or DevOps engineers. However, it is always good to be aware of their operations.

Other AWS network services

There are lots of other AWS Network Services that we do not have the space to cover. Some of the more important ones are listed in this section:

- **AWS Transit VPC** (https://aws.amazon.com/blogs/aws/aws-solution-transit-vpc/): This is a way to connect multiple virtual private clouds to a common VPC that serves as a transit center. This is a relatively new service, but it can minimize the connection that you need to set up and manage. This can also serve as a tool when you need to share resources between separate AWS accounts.
- **Amazon GuardDuty** (https://aws.amazon.com/guardduty/): This is a managed threat detection service that continuously monitors for malicious or unauthorized behavior to help protect our AWS workloads. It monitors API calls or potentially unauthorized deployments.
- **AWS WAF** (https://aws.amazon.com/waf/): This is a web application firewall that helps protect web applications from common exploits. We can define customized web security rules to allow or block web traffic.
- **AWS Shield** (https://aws.amazon.com/shield/): This is a managed **Distributed Denial of Service** (**DDoS**) protection service that safeguards applications running on AWS. The protection service is free for all customers at the basic level; the advanced version of AWS Shield is a fee-based service.

Summary

In this chapter, we looked at AWS Cloud Networking services. We went over the AWS network definitions of Region, Availability Zone, Edge Locations, and Transit Center. By understanding the overall AWS network, this gives us a good idea of some of the limitations and contains for the other AWS network services. Throughout this chapter, we used the AWS CLI, the Python Boto3 library, as well as CloudFormation to automate some of the tasks.

We covered the AWS virtual private cloud in depth with the configuration of the route table and route targets. The example on security groups and network ACL controls the security for our VPC. We also looked at Elastic IP and NAT Gateways regarding allowing external access.

There are two ways to connect AWS VPC to on-premise networks: Direct Connect and IPSec VPN. We briefly looked at each and the advantages of using them. Toward the end of this chapter, we looked at network scaling services offered by AWS, including Elastic Load Balancing, Route53 DNS, and CloudFront.

In `Chapter 11`, *Working with Git*, we will take a more in-depth look at the version control system we have been working with: Git.

11
Working with Git

We have worked on various aspects of network automation with Python, Ansible, and many other tools. If you have been following along with the examples, in the first nine chapters of the book, we have used over 150 files containing over 5,300 lines of code. That's pretty good for network engineers who may have been working primarily with the command-line interface! With our new set of scripts and tools, we are now ready to go out and conquer our network tasks, right? Well, not so fast, my fellow network ninjas.

The first task we face with the code files is how to keep them in a location where they can be retrieved and used by us and others. Ideally, this location would be the only place where the latest version of the file is kept. After the initial release, we might add features and fix bugs in the future, so we would like a way to track these changes and keep the latest ones available for download. If the new changes do not work, we would like to rollback the changes and reflect the differences in the history of the file. This would give us a good idea of the evolution of the code files.

The second question is the collaboration process between our team members. If we work with other network engineers, we will need to work collectively on the files. The files can be the Python scripts, Ansible Playbook, Jinja2 templates, INI-style configuration files, and many others. The point is any kind of text-based files should be tracked with multiple input that everybody in the team should be able to see.

The third question is accountability. Once we have a system that allows for multiple inputs and changes, we need to mark these changes with an appropriate track record to reflect the owner of the change. The track record should also include a brief reason for the change so the person reviewing the history can get an understanding of why the change was made.

Working with Git

These are some of the main challenges a version-control (or source-control) system tries to solve. To be fair, version control can exist in forms other than a dedicated system. For example, if I open up my Microsoft Word program, the file constantly saves itself, and I can go back in time to revisit the changes or rollback to a previous version. The version-control system we are focused on here is standalone software tools with the primary purpose of tracking software changes.

There is no shortage of different source-control tools in software engineering, both proprietary and open source. Some of the more popular open source version-control systems are CVS, SVN, Mercurial, and Git. In this chapter, we will focus on the source-control system **Git**, the tool that we have been downloading in many of the `.software` packages we have used in this book. We will be taking a more in-depth look at the tool. Git is the de facto version-control system for many large, open source projects, including Python and the Linux kernel.

 As of February 2017, the CPython development process has moved to GitHub. It was a work in progress since January 2015. For more information, check out PEP 512 at `https://www.python.org/dev/peps/pep-0512/`.

Before we dive into the working examples of Git, let's take a look at the history and advantages of the Git system.

Introduction to Git

Git was created by Linus Torvalds, the creator of the Linux kernel, in April 2005. With his dry wit, he has affectionately called the tool the information manager from hell. In an interview with the Linux Foundation, Linus mentioned that he felt source-control management was just about the least interesting thing in the computing world (`https://www.linuxfoundation.org/blog/10-years-of-git-an-interview-with-git-creator-linus-torvalds/`). Nevertheless, he created the tool after a disagreement between the Linux kernel developer community and BitKeeper, the proprietary system they were using at the time.

 What does the name Git stand for? In British English slang, a Git is an insult denoting an unpleasant, annoying, childish person. With his dry humor, Linus said he is an egotistical bastard and that he named all of his projects after himself. First Linux, now Git. However, some suggested that the name is short for **Global Information Tracker** (**GIT**). You can be the judge.

The project came together really quickly. About ten days after its creation (yeah, you read that right), Linus felt the basic ideas for Git were right and started to commit the first Linux kernel code with Git. The rest, as they say, is history. More than ten years after its creation, it is still meeting all the expectations of the Linux kernel project. It took over as the version-control system for many other open source projects despite the inherent inertia in switching source-control systems. After many years of hosting the Python code from Mercurial at `https://hg.python.org/`, the project was switched to Git on GitHub in February of 2017.

Benefits of Git

The success of hosting large and distributed open source projects, such as the Linux kernel and Python, speaks to the advantages of Git. This is especially significant given that Git is a relatively new source-control tool and people do not tend to switch to a new tool unless it offers significant advantages over the old tool. Let's look at some of the benefits of Git:

- **Distributed development**: Git supports parallel, independent, and simultaneous development in private repositories offline. Compare this to some other version-control systems that require constant synchronization with a central repository; this allows significantly greater flexibility for the developers.
- **Scale to handle thousands of developers**: The number of developers working on different parts of some of the open source projects is in the thousands. Git supports the integration of their work reliably.
- **Performance**: Linus was determined to make sure Git was fast and efficient. To save space and transfer time for the sheer volume of updates for the Linux kernel code alone, compression and a delta check would be needed to make Git fast and efficient.
- **Accountability and immutability**: Git enforces a change log on every commit that changes a file so that there is a trail for all the changes and the reason behind them. The data objects in Git cannot be modified after they were created and placed in the database, making them immutable. This further enforces accountability.
- **Atomic transactions**: The integrity of the repository is ensured as the different, but related, change is performed either all together or not at all. This will ensure the repository is not left in a partially-changed or corrupted state.
- **Complete repositories**: Each repository has a complete copy of all historical revisions of every file.
- **Free, as in freedom**: The origin of the Git tool was born out of the disagreement between free, as in beer version of the Linux kernel with BitKeeper VCS, it makes sense that the tool has a very liberal usage license.

Let's take a look at some of the terms used in Git.

Git terminology

Here are some Git terminologies we should be familiar with:

- **Ref**: The name that begins with `refs` that point to an object.
- **Repository**: A database that contains all of a project's information, files, metadata, and history. It contains a collection of `ref` for all the collections of objects.
- **Branch**: An active line of development. The most recent commit is the `tip` or the `HEAD` of that branch. A repository can have multiple branches, but your `working tree` or `working directory` can only be associated with one branch. This is sometimes referred to as the current or `checked out` branch.
- **Checkout**: The action of updating all or part of the working tree to a particular point.
- **Commit**: A point in time in Git history, or it can mean to store a new snapshot into the repository.
- **Merge**: The action to bring the content of another branch into the current branch. For example, I am merging the `development` branch with the `master` branch.
- **Fetch**: The action of getting the content from a remote repository.
- **Pull**: Fetching and merging a repository.
- **Tag**: A mark in a point in time in a repository that is significant. In `Chapter 4`, *The Python Automation Framework – Ansible Basics*, we saw the tag used to specify the release points, `v2.5.0a1`.

This is not a complete list; please refer to the Git glossary, `https://git-scm.com/docs/gitglossary`, for more terms and their definitions.

Git and GitHub

Git and GitHub are not the same thing. Sometimes, for engineers who are new to version-control systems, this is confusing. Git is a revision-control system while GitHub, `https://github.com/`, is a centralized hosting service for Git repositories.

Chapter 11

Because Git is a decentralized system, GitHub stores a copy of our project's repository, just like any other developer. Often, we just designate the GitHub repository as the project's central repository and all other developers push and pull their changes to and from that repository.

GitHub takes this idea of being the centralized repository in a distributed system further by using the `fork` and `pull requests` mechanisms. For projects hosted on GitHub, encourage developers to `fork` the repository, or make a copy of the repository, and work on that copy as their centralized repository. After making changes, they can send a `pull request` to the main project, and the project maintainers can review the changes and `commit` the changes if they see fit. GitHub also adds the web interface to the repositories besides command line; this makes Git more user-friendly.

Setting up Git

So far, we have been using Git to just download files from GitHub. In this section, we will go a bit further by setting up Git variables so we can start committing our files. I am going to use the same Ubuntu 16.04 host in the example. The installation process is well-documented; if you are using a different version of Linux or other operating systems, a quick search should land you at the right set of instructions.

If you have not done so already, install Git via the `apt` package-management tool:

```
$ sudo apt-get update
$ sudo apt-get install -y git
$ git --version
git version 2.7.4
```

Once `git` is installed, we need to configure a few things so our commit messages can contain the correct information:

```
$ git config --global user.name "Your Name"
$ git config --global user.email "email@domain.com"
$ git config --list
user.name=Your Name
user.email=email@domain.com
```

Alternatively, you can modify the information in the `~/.gitconfig` file:

```
$ cat ~/.gitconfig
[user]
  name = Your Name
  email = email@domain.com
```

[355]

Working with Git

There are many other options in Git that we can change, but the name and email are the ones that allow us to commit the change without getting a warning. Personally, I like to use VIM, instead of the default Emac, as my text editor for typing commit messages:

```
(optional)
$ git config --global core.editor "vim"
$ git config --list
user.name=Your Name
user.email=email@domain.com
core.editor=vim
```

Before we move on to using Git, let's go over the idea of a `gitignore` file.

Gitignore

From time to time, there are files you do not want Git to check into GitHub or other repositories. The easiest way to do this is to create `.gitignore` in the `repository` folder; Git will use it to determine which files a directory should ignore before you make a commit. This file should be committed into the repository to share the ignore rules with other users.

This file can include language-specific files, for example, let's exclude the Python `Byte-compiled` files:

```
# Byte-compiled / optimized / DLL files
__pycache__/
*.py[cod]
*$py.class
```

We can also include files that are specific to your operating system:

```
# OSX
# =========================

.DS_Store
.AppleDouble
.LSOverride
```

You can learn more about `.gitignore` on GitHub's help page: `https://help.github.com/articles/ignoring-files/`. Here are some other references:

- Gitignore manual: `https://git-scm.com/docs/gitignore`
- GitHub's collection of `.gitignore` templates: `https://github.com/github/gitignore`

- Python language .gitignore example: https://github.com/github/gitignore/blob/master/Python.gitignore
- The .gitignore file for this book's repository: https://github.com/PacktPublishing/Mastering-Python-Networking-Second-Edition/blob/master/.gitignore

I see the .gitignore file as a file that should be created at the same time as any new repository. That is why this concept is introduced as early as possible. We will take a look at some of the Git usage examples in the next section.

Git usage examples

Most of the time, when we work with Git, we will use the command line:

```
$ git --help
usage: git [--version] [--help] [-C <path>] [-c name=value]
           [--exec-path[=<path>]] [--html-path] [--man-path] [--info-path]
           [-p | --paginate | --no-pager] [--no-replace-objects] [--bare]
           [--git-dir=<path>] [--work-tree=<path>] [--namespace=<name>]
           <command> [<args>]
```

We will create a `repository` and create a file inside the repository:

```
$ mkdir TestRepo
$ cd TestRepo/
$ git init
Initialized empty Git repository in /home/echou/Master_Python_Networking_second_edition/Chapter11/TestRepo/.git/
$ echo "this is my test file" > myFile.txt
```

When the repository was initialized with Git, a new hidden folder of .git was added to the directory. It contains all the Git-related files:

```
$ ls -a
.  ..  .git  myFile.txt

$ ls .git/
branches  config  description  HEAD  hooks  info  objects  refs
```

There are several locations Git receives its configurations in a hierarchy format. You can use the `git config -l` command to see the aggregated configuration:

```
$ ls .git/config
.git/config

$ ls ~/.gitconfig
/home/echou/.gitconfig

$ git config -l
user.name=Eric Chou
user.email=<email>
core.editor=vim
core.repositoryformatversion=0
core.filemode=true
core.bare=false
core.logallrefupdates=true
```

When we create a file in the repository, it is not tracked. For `git` to be aware of the file, we need to add the file:

```
$ git status
On branch master

Initial commit

Untracked files:
  (use "git add <file>..." to include in what will be committed)

    myFile.txt

nothing added to commit but untracked files present (use "git add" to track)

$ git add myFile.txt
$ git status
On branch master

Initial commit

Changes to be committed:
  (use "git rm --cached <file>..." to unstage)

    new file:   myFile.txt
```

When you add the file, it is in a staged status. To make the changes official, we will need to commit the change:

```
$ git commit -m "adding myFile.txt"
[master (root-commit) 5f579ab] adding myFile.txt
 1 file changed, 1 insertion(+)
 create mode 100644 myFile.txt

$ git status
On branch master
nothing to commit, working directory clean
```

In the last example, we provided the commit message with the -m option when we issue the commit statement. If we did not use the option, we would have been taken to a page to provide the commit message. In our scenario, we configured the text editor to be vim so we will be able to use vim to edit the message.

Let's make some changes to the file and commit it:

```
$ vim myFile.txt
$ cat myFile.txt
this is the second iteration of my test file
$ git status
On branch master
Changes not staged for commit:
  (use "git add <file>..." to update what will be committed)
  (use "git checkout -- <file>..." to discard changes in working directory)

    modified:   myFile.txt
$ git add myFile.txt
$ git commit -m "made modificaitons to myFile.txt"
[master a3dd3ea] made modificaitons to myFile.txt
 1 file changed, 1 insertion(+), 1 deletion(-)
```

The `git commit` number is a `SHA1 hash`, which an important feature. If we had followed the same step on another computer, our `SHA1 hash` value would be the same. This is how Git knows the two repositories are identical even when they are worked on in parallel.

We can show the history of the commits with `git log`. The entries are shown in reverse chronological order; each commit shows the author's name and email address, the date, the log message, as well as the internal identification number of the commit:

```
$ git log
commit a3dd3ea8e6eb15b57d1f390ce0d2c3a03f07a038
Author: Eric Chou <echou@yahoo.com>
Date:   Fri Jul 20 09:58:24 2018 -0700
```

```
        made modificaitons to myFile.txt

commit 5f579ab1e9a3fae13aa7f1b8092055213157524d
Author: Eric Chou <echou@yahoo.com>
Date:   Fri Jul 20 08:05:09 2018 -0700

        adding myFile.txt
```

We can also show more details about the change using the commit ID:

```
$ git show a3dd3ea8e6eb15b57d1f390ce0d2c3a03f07a038
commit a3dd3ea8e6eb15b57d1f390ce0d2c3a03f07a038
Author: Eric Chou <echou@yahoo.com>
Date:   Fri Jul 20 09:58:24 2018 -0700

        made modificaitons to myFile.txt

diff --git a/myFile.txt b/myFile.txt
index 6ccb42e..69e7d47 100644
--- a/myFile.txt
+++ b/myFile.txt
@@ -1 +1 @@
-this is my test file
+this is the second iteration of my test file
```

If you need to revert the changes you have made, you can choose between `revert` and `reset`. Revert changes all the file for a specific commit back to their state before the commit:

```
$ git revert a3dd3ea8e6eb15b57d1f390ce0d2c3a03f07a038
[master 9818f29] Revert "made modificaitons to myFile.txt"
 1 file changed, 1 insertion(+), 1 deletion(-)

# Check to verified the file content was before the second change.
$ cat myFile.txt
this is my test file
```

The `revert` command will keep the commit you reverted and make a new commit. You will be able to see all the changes up to that point, including the revert:

```
$ git log
commit 9818f298f477fd880db6cb87112b50edc392f7fa
Author: Eric Chou <echou@yahoo.com>
Date:   Fri Jul 20 13:11:30 2018 -0700

        Revert "made modificaitons to myFile.txt"

        This reverts commit a3dd3ea8e6eb15b57d1f390ce0d2c3a03f07a038.
```

```
        modified:   reverted the change to myFile.txt

commit a3dd3ea8e6eb15b57d1f390ce0d2c3a03f07a038
Author: Eric Chou <echou@yahoo.com>
Date:   Fri Jul 20 09:58:24 2018 -0700

    made modificaitons to myFile.txt

commit 5f579ab1e9a3fae13aa7f1b8092055213157524d
Author: Eric Chou <echou@yahoo.com>
Date:   Fri Jul 20 08:05:09 2018 -0700

    adding myFile.txt
```

The `reset` option will reset the status of your repository to an older version and discard all the changes in between:

```
$ git reset --hard a3dd3ea8e6eb15b57d1f390ce0d2c3a03f07a038
HEAD is now at a3dd3ea made modificaitons to myFile.txt

$ git log
commit a3dd3ea8e6eb15b57d1f390ce0d2c3a03f07a038
Author: Eric Chou <echou@yahoo.com>
Date:   Fri Jul 20 09:58:24 2018 -0700

    made modificaitons to myFile.txt

commit 5f579ab1e9a3fae13aa7f1b8092055213157524d
Author: Eric Chou <echou@yahoo.com>
Date:   Fri Jul 20 08:05:09 2018 -0700

    adding myFile.txt
```

Personally, I like to keep all the history, including any rollbacks that I have done. Therefore, when I need to rollback a change, I usually pick `revert` instead of `reset`.

A `branch` in `git` is a line of development within a repository. Git allows many branches and thus different lines of development within a repository. By default, we have the master branch. There are many reasons for branching, but most of them represent an individual customer release or a development phase, that is, the `dev` branch. Let's create a `dev` branch within our repository:

```
$ git branch dev
$ git branch
  dev
* master
```

Working with Git

To start working on the branch, we will need to `checkout` the branch:

```
$ git checkout dev
Switched to branch 'dev'
$ git branch
* dev
  master
```

Let's add a second file to the `dev` branch:

```
$ echo "my second file" > mySecondFile.txt
$ git add mySecondFile.txt
$ git commit -m "added mySecondFile.txt to dev branch"
[dev c983730] added mySecondFile.txt to dev branch
 1 file changed, 1 insertion(+)
 create mode 100644 mySecondFile.txt
```

We can go back to the `master` branch and verify that the two lines of development are separate:

```
$ git branch
* dev
  master
$ git checkout master
Switched to branch 'master'
$ ls
myFile.txt
$ git checkout dev
Switched to branch 'dev'
$ ls
myFile.txt mySecondFile.txt
```

To have the contents in the `dev` branch be written into the `master` branch, we will need to `merge` them:

```
$ git branch
* dev
  master
$ git checkout master
$ git merge dev master
Updating a3dd3ea..c983730
Fast-forward
 mySecondFile.txt | 1 +
 1 file changed, 1 insertion(+)
 create mode 100644 mySecondFile.txt
$ git branch
  dev
* master
```

[362]

```
$ ls
myFile.txt mySecondFile.txt
```

We can use `git rm` to remove a file. Let's create a third file and remove it:

```
$ touch myThirdFile.txt
$ git add myThirdFile.txt
$ git commit -m "adding myThirdFile.txt"
[master 2ec5f7d] adding myThirdFile.txt
 1 file changed, 0 insertions(+), 0 deletions(-)
 create mode 100644 myThirdFile.txt
$ ls
myFile.txt mySecondFile.txt myThirdFile.txt
$ git rm myThirdFile.txt
rm 'myThirdFile.txt'
$ git status
On branch master
Changes to be committed:
  (use "git reset HEAD <file>..." to unstage)

    deleted:    myThirdFile.txt
$ git commit -m "deleted myThirdFile.txt"
[master bc078a9] deleted myThirdFile.txt
 1 file changed, 0 insertions(+), 0 deletions(-)
 delete mode 100644 myThirdFile.txt
```

We will be able to see the last two changes in the log:

```
$ git log
commit bc078a97e41d1614c1ba1f81f72acbcd95c0728c
Author: Eric Chou <echou@yahoo.com>
Date: Fri Jul 20 14:02:02 2018 -0700

    deleted myThirdFile.txt

commit 2ec5f7d1a734b2cc74343ce45075917b79cc7293
Author: Eric Chou <echou@yahoo.com>
Date: Fri Jul 20 14:01:18 2018 -0700

    adding myThirdFile.txt
```

We have gone through most of the basic operations we would use for Git. Let's take a look at how to use GitHub to share our repository.

GitHub example

In this example, we will use GitHub as the centralized location to synchronize our local repository and share with other users.

We will create a repository on GitHub. By default, GitHub has a free public repository; in my case, I pay a small monthly fee to host private repositories. At the time of creation, you can choose to create the license and the .gitignore file:

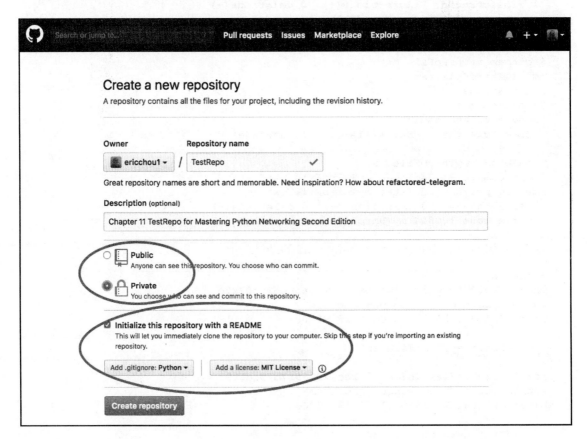

GitHub private repository

Once the repository is created, we can find the URL for this repository:

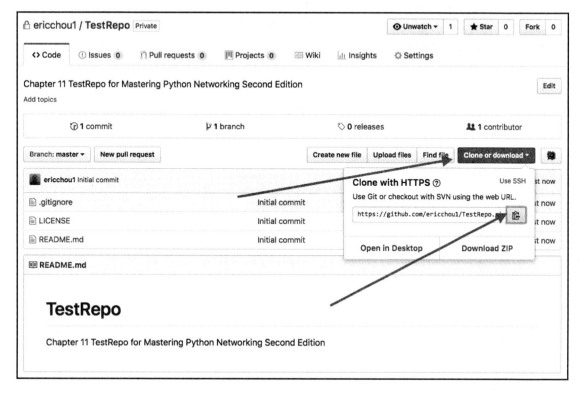

GitHub repository URL

We will use this URL to create a `remote` target; we will name it `gitHubRepo`:

```
$ git remote add gitHubRepo https://github.com/ericchou1/TestRepo.git
$ git remote -v
gitHubRepo https://github.com/ericchou1/TestRepo.git (fetch)
gitHubRepo https://github.com/ericchou1/TestRepo.git (push)
```

Since we chose to create a `README.md` and `LICENSE` file during creation, the remote repository and current repository are not the same. If we were to push local changes to the GitHub repository, we would receive the following error:

```
$ git push gitHubRepo master
Username for 'https://github.com': echou@yahoo.com
Password for 'https://echou@yahoo.com@github.com':
To https://github.com/ericchou1/TestRepo.git
 ! [rejected]        master -> master (fetch first)
```

We will go ahead and use `git pull` to get the new files from GitHub:

```
$ git pull gitHubRepo master
Username for 'https://github.com': <username>
Password for 'https://<username>@github.com':
From https://github.com/ericchou1/TestRepo
 * branch            master     -> FETCH_HEAD
Merge made by the 'recursive' strategy.
 .gitignore | 104 ++++++++++++++++++++++++++++++++++++++++++++++++++++++++++++++++
 LICENSE    |  21 ++++++++++++
 README.md  |   2 ++
 3 files changed, 127 insertions(+)
 create mode 100644 .gitignore
 create mode 100644 LICENSE
 create mode 100644 README.md
```

Now we will be able to `push` the contents over to GitHub:

```
$ git push gitHubRepo master
Username for 'https://github.com': <username>
Password for 'https://<username>@github.com':
Counting objects: 15, done.
Compressing objects: 100% (9/9), done.
Writing objects: 100% (15/15), 1.51 KiB | 0 bytes/s, done.
Total 15 (delta 1), reused 0 (delta 0)
remote: Resolving deltas: 100% (1/1), done.
To https://github.com/ericchou1/TestRepo.git
   a001b81..0aa362a  master -> master
```

We can verify the content of the GitHub repository on the web page:

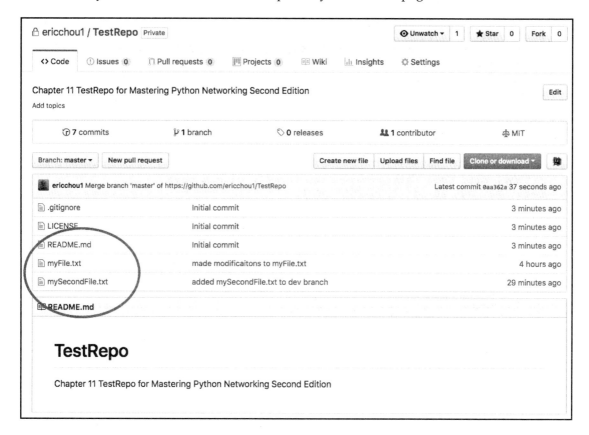

GitHub repository

Now another user can simply make a copy, or `clone`, of the repository:

```
[This is operated from another host]
$ cd /tmp
$ git clone https://github.com/ericchou1/TestRepo.git
Cloning into 'TestRepo'...
remote: Counting objects: 20, done.
remote: Compressing objects: 100% (13/13), done.
remote: Total 20 (delta 2), reused 15 (delta 1), pack-reused 0
Unpacking objects: 100% (20/20), done.
$ cd TestRepo/
$ ls
LICENSE   myFile.txt
README.md mySecondFile.txt
```

Working with Git

This copied repository will be the exact copy of my original repository, including all the commit history:

```
$ git log
commit 0aa362a47782e7714ca946ba852f395083116ce5 (HEAD -> master,
origin/master, origin/HEAD)
Merge: bc078a9 a001b81
Author: Eric Chou <echou@yahoo.com>
Date: Fri Jul 20 14:18:58 2018 -0700

    Merge branch 'master' of https://github.com/ericchou1/TestRepo

commit a001b816bb75c63237cbc93067dffcc573c05aa2
Author: Eric Chou <ericchou1@users.noreply.github.com>
Date: Fri Jul 20 14:16:30 2018 -0700

    Initial commit
...
```

I can also invite another person as a collaborator for the project under the repository setting:

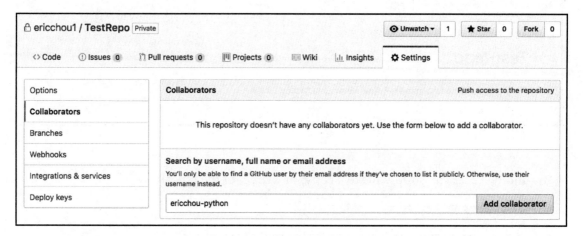

Repository invite

In the next example, we will see how we can fork a repository and perform a pull request for a repository that we do not maintain.

Collaborating with pull requests

As mentioned, Git supports collaboration between developers for a single project. We will take a look at how it is done when the code is hosted on GitHub.

In this case, I am going to take a look at the GitHub repository for this book. I am going to use a different GitHub handle, so I appear as a different user. I will click on the **Fork** bottom to make a copy of the repository in my personal account:

Git fork bottom

It will take a few seconds to make a copy:

Git fork in progress

Working with Git

After it is forked, we will have a copy of the repository in our personal account:

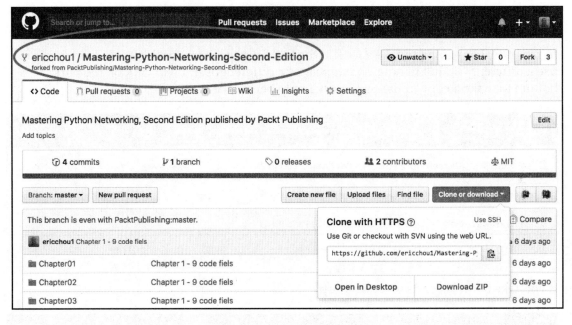

Git fork

We can follow the same steps we have used before to make some modifications to the files. In this case, I will make some changes to the README.md file. After the change is made, I can click on the **New pull request** button to create a pull request:

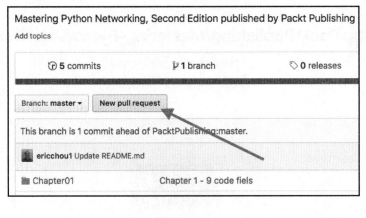

Pull request

When making a pull request, we should fill in as much information as possible to provide justifications for making the change:

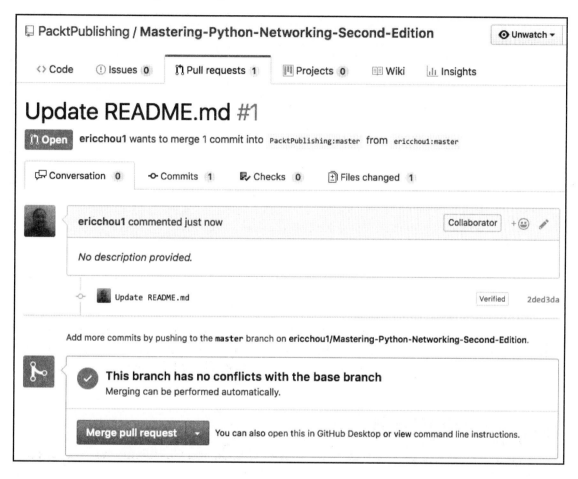

Pull request details

Working with Git

The repository maintainer will receive a notification of the pull request; if accepted, the change will make its way to the original repository:

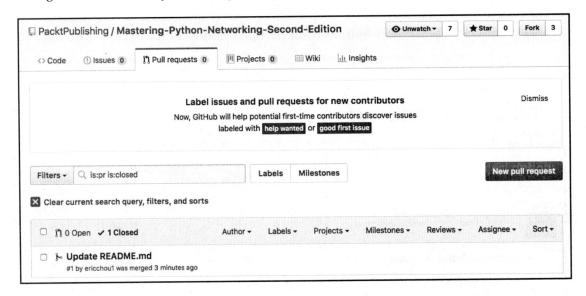

Pull request record

GitHub provides an excellent platform for collaboration with other developers; this is quickly becoming the de facto development choice for many large, open source projects. In the following section, let's take a look at how we can use Git with Python.

Git with Python

There are some Python packages that we can use with Git and GitHub. In this section, we will take a look at the GitPython and PyGithub libraries.

GitPython

We can use the GitPython package, `https://gitpython.readthedocs.io/en/stable/index.html`, to work with our Git repository. We will install the package and use the Python shell to construct a `Repo` object. From there, we can list all the commits in the repository:

```
$ sudo pip3 install gitpython
$ python3
```

```
>>> from git import Repo
>>> repo =
Repo('/home/echou/Master_Python_Networking_second_edition/Chapter11/TestRep
o')
>>> for commits in list(repo.iter_commits('master')):
...     print(commits)
...
0aa362a47782e7714ca946ba852f395083116ce5
a001b816bb75c63237cbc93067dffcc573c05aa2
bc078a97e41d1614c1ba1f81f72acbcd95c0728c
2ec5f7d1a734b2cc74343ce45075917b79cc7293
c98373069f27d8b98d1ddacffe51b8fa7a30cf28
a3dd3ea8e6eb15b57d1f390ce0d2c3a03f07a038
5f579ab1e9a3fae13aa7f1b8092055213157524d
```

We can also look at the index entries:

```
>>> for (path, stage), entry in index.entries.items():
...     print(path, stage, entry)
...
mySecondFile.txt 0 100644 75d6370ae31008f683cf18ed086098d05bf0e4dc 0
mySecondFile.txt
LICENSE 0 100644 52feb16b34de141a7567e4d18164fe2400e9229a 0 LICENSE
myFile.txt 0 100644 69e7d4728965c885180315c0d4c206637b3f6bad 0 myFile.txt
.gitignore 0 100644 894a44cc066a027465cd26d634948d56d13af9af 0 .gitignore
README.md 0 100644 a29fe688a14d119c20790195a815d078976c3bc6 0 README.md
>>>
```

GitPython offers good integration with all the Git functions. However, it is not the easiest to work with. We need to understand the terms and structure of Git to take full advantage of GitPython. But it is always good to keep it in mind in case we need it for other projects.

PyGitHub

Let's look at using the PyGitHub package, http://pygithub.readthedocs.io/en/latest/, to interact with GitHub repositories. The package is a wrapper around GitHub APIv3, https://developer.github.com/v3/:

```
$ sudo pip install pygithub
$ sudo pip3 install pygithub
```

Working with Git

Let's use the Python shell to print out the user's current repository:

```
$ python3
>>> from github import Github
>>> g = Github("ericchou1", "<password>")
>>> for repo in g.get_user().get_repos():
...     print(repo.name)
...
ansible
...
-Hands-on-Network-Programming-with-Python
Mastering-Python-Networking
Mastering-Python-Networking-Second-Edition
>>>
```

For more programmatic access, we can also create more granular control using an access token. Github allows a token to be associated with the selected rights:

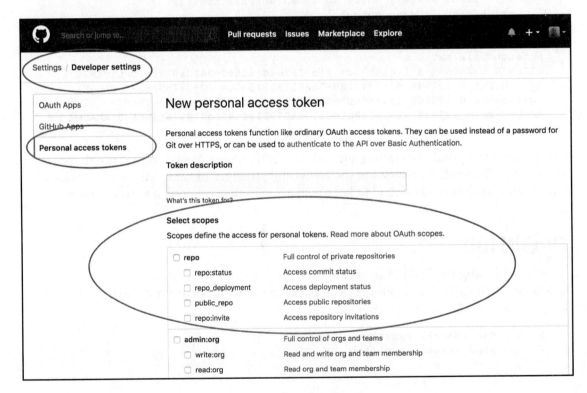

GitHub token generation

The output is a bit different if you use the access token as the authentication mechanism:

```
>>> from github import Github
>>> g = Github("<token>")
>>> for repo in g.get_user().get_repos():
...     print(repo)
...
Repository(full_name="oreillymedia/distributed_denial_of_service_ddos")
Repository(full_name="PacktPublishing/-Hands-on-Network-Programming-with-Python")
Repository(full_name="PacktPublishing/Mastering-Python-Networking")
Repository(full_name="PacktPublishing/Mastering-Python-Networking-Second-Edition")
...
```

Now that we are familiar with Git, GitHub, and some of the Python packages, we can use them to work with the technology. We will take a look at some practical examples in the coming section.

Automating configuration backup

In this example, we will use PyGithub to back up a directory containing our router configurations. We have seen how we can retrieve the information from our devices with Python or Ansible; we can now check them into GitHub.

We have a subdirectory, named `config`, with our router configs in text format:

```
$ ls configs/
iosv-1 iosv-2

$ cat configs/iosv-1
Building configuration...

Current configuration : 4573 bytes
!
! Last configuration change at 02:50:05 UTC Sat Jun 2 2018 by cisco
!
version 15.6
service timestamps debug datetime msec
...
```

We can use the following script to retrieve the latest index from our GitHub repository, build the content that we need to commit, and automatically commit the configuration:

```
$ cat Chapter11_1.py
#!/usr/bin/env python3
# reference:
https://stackoverflow.com/questions/38594717/how-do-i-push-new-files-to-github

from github import Github, InputGitTreeElement
import os

github_token = '<token>'
configs_dir = 'configs'
github_repo = 'TestRepo'

# Retrieve the list of files in configs directory
file_list = []
for dirpath, dirname, filenames in os.walk(configs_dir):
    for f in filenames:
        file_list.append(configs_dir + "/" + f)

g = Github(github_token)
repo = g.get_user().get_repo(github_repo)

commit_message = 'add configs'
master_ref = repo.get_git_ref('heads/master')
master_sha = master_ref.object.sha
base_tree = repo.get_git_tree(master_sha)

element_list = list()

for entry in file_list:
    with open(entry, 'r') as input_file:
        data = input_file.read()
    element = InputGitTreeElement(entry, '100644', 'blob', data)
    element_list.append(element)

# Create tree and commit
tree = repo.create_git_tree(element_list, base_tree)
parent = repo.get_git_commit(master_sha)
commit = repo.create_git_commit(commit_message, tree, [parent])
master_ref.edit(commit.sha)
```

Chapter 11

We can see the `configs` directory in the GitHub repository:

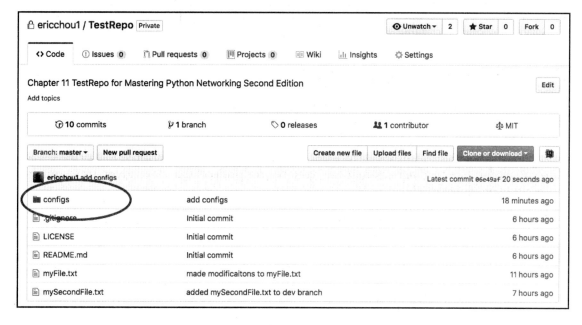

Configs directory

The commit history shows the commit from our script:

Commit history

In the *GitHub example* section, we saw how we could collaborate with other developers by forking the repository and making pull requests. Let's look at how we can further collaborate with Git.

Collaborating with Git

Git is an awesome collaboration technology, and GitHub is an incredibly effective way to develop projects together. GitHub provides a place for anyone in the world with internet access to share their thoughts and code for free. We know how to use Git and some of the basic collaboration steps using GitHub, but how do we join and contribute to a project? Sure, we would like to give back to these open source projects that have given us so much, but how do we get started?

In this section, we'll look at some of the things to know about software-development collaboration using Git and GitHub:

- **Start small**: One of the most important things to understand is the role we can play within a team. We might be awesome at network engineering but a mediocre Python developer. There are plenty of things we can do that don't involve being a highly-skilled developer. Don't be afraid to start small, documentation and testing are two good ways to get your foot in the door as a contributor.
- **Learn the ecosystem**: With any project, large or small, there is a set of conventions and a culture that has been established. We are all drawn to Python for its easy-to-read syntax and beginner-friendly culture; they also have a development guide that is centered around that ideology (https://devguide.python.org/). The Ansible project, on the other hand, also has an extensive community guide (https://docs.ansible.com/ansible/latest/community/index.html). It includes the code of conduct, the pull request process, how to report bugs, and the release process. Read these guides and learn the ecosystem for the project of interest.
- **Make a branch**: I have made the mistake of forking a project and making a pull request for the main branch. The main branch should be left alone for the core contributors to make changes to. We should create a separate branch for our contribution and allow the branch to be merged at a later date.

- **Keep forked repository synchronized**: Once you have forked a project, there is no rule that forces the cloned repository to sync with the main repository. We should make a point to regularly do `git pull` (get the code and merge locally) or `git fetch` (get the code with any change locally) to make sure we have the latest copy of the main repository.
- **Be friendly**: Just as in the real world, the virtual world has no place for hostility. When discussing an issue, be civil and friendly, even in disagreements.

Git and GitHub provide a way for any motivated individual to make a difference by making it easy to collaborate on projects. We are all empowered to contribute to any open source or private projects that we find interesting.

Summary

In this chapter, we looked at the version-control system known as Git and its close sibling, GitHub. Git was developed by Linus Torvolds in 2005 to help develop the Linux kernel and later adopted by other open source projects as the source-control system. Git is a fast, distributed, and scalable system. GitHub provides a centralized location to host Git repositories on the internet that allow anybody with an internet connection to collaborate.

We looked at how to use Git in the command line, its various operations, and how they are applied in GitHub. We also studied two of the popular Python libraries for working with Git: GitPython and PyGitHub. We ended the chapter with a configuration backup example and notes about project collaboration.

In `Chapter 12`, *Continuous Integration with Jenkins*, we will look at another popular open source tool used for continuous integration and deployment: Jenkins.

12
Continuous Integration with Jenkins

The network touches every part of the technology stack; in all of the environments I have worked in, it is always a Tier-Zero service. It is a foundation service that other services rely on for their services to work. In the minds of other engineers, business managers, operators, and support staff, the network should just work. It should always be accessible and function correctly—a good network is a network that nobody hears about.

Of course, as network engineers, we know the network is as complex as any other technology stack. Due to its complexity, the constructs that make up a running network can be fragile at times. Sometimes, I look at a network and wonder how it can work at all, let alone how it's been running for months and years without business impacts.

Part of the reason we are interested in network automation is to find ways to repeat our network-change process reliably and consistently. By using Python scripts or the Ansible framework, we can make sure the change that we make will stay consistent and be reliably applied. As we saw in the last chapter, we can use Git and GitHub to store components of the process, such as templates, scripts, requirements, and files, reliably. The code that makes up the infrastructure is version-controlled, collaborated, and accountable for changes. But how do we tie all the pieces together? In this chapter, we will look at a popular open source tool that can optimize the network-management pipeline, called Jenkins.

Traditional change-management process

For engineers who have worked in a large network environment, they know the impact of a network change gone wrong can be big. We can make hundreds of changes without any issues, but all it takes is one bad change that can cause the network to have a negative impact on the business.

There is no shortage of war stories about network outages causing business pain. One of the most visible and large-scale AWS EC2 outage in 2011 was caused by a network change that was part of our normal AWS scaling activities in the AWS US-East region. The change occurred at 00:47 PDT and caused a brown-out for various services for over 12 hours, losing millions of dollars for Amazon in the process. More importantly, the reputation of the relatively young service took a serious hit. IT decision makers will point to the outage as reasons to NOT migrate to AWS cloud. It took many years to rebuild its reputation. You can read more about the incident report at https://aws.amazon.com/message/65648/.

Due to its potential impact and complexity, in many environments, the **Change-Advisory Board (CAB)** is implemented for networks. The typical CAB process is as follows:

1. The network engineer will design the change and write out the detail steps required of the change. This can include the reason for the change, the devices involved, the commands that will be applied or deleted, how to verify the output, and the expected outcome for each of the steps.
2. The network engineer is typically required to ask for a technical review from a peer first. Depending on the nature of the change, there can be different levels of peer review. The simple changes can require a single peer technical review; the complex change might require a senior designated engineer for approval.
3. The CAB meeting is generally scheduled for set times with emergency ad-hoc meetings available.
4. The engineer will present the change to the board. The board will ask the necessary questions, assess the impact, and either approve or deny the change request.
5. The change will be carried out, either by the original engineer or another engineer, at the scheduled change window.

This process sounds reasonable and inclusive but proves to have a few challenges in practice:

- **Write-ups are time-consuming**: It typically takes a lot of time for the design engineer to write up the document, and sometimes the writing process takes longer than the time to apply the change. This is generally due to the fact that all network changes are potentially impactful and we need to document the process for both technical and non-technical CAB members.

- **Engineer expertise**: There are different levels of engineering expertise, some are more experienced, and they are typically the most sought-after resources. We should reserve their time for tackling the most complex network issues, not reviewing basic network changes.
- **Meetings are time-consuming**: It takes a lot of effort to put together meetings and have each member show up. What happens if a required approval person is on vacation or sick? What if you need the network change to be made prior to the scheduled CAB time?

These are just some of the bigger challenges of the human-based CAB process. Personally, I hate the CAB process with a passion. I do not dispute the need for peer review and prioritization; however, I think we need to minimize the potential overhead involved. Let's look at a potential pipeline that has been adopted in the software-engineering pipeline.

Introduction to continuous integration

Continuous Integration (CI) in software development is a way to publish small changes to the code base quickly, in the context of tests and validation built-in. The keys are to classify the changes to be CI-compatible, that is, not overly complex, and small enough to be applied that they can be backed out easily. The tests and validation process is built in an automated way to gain a baseline of confidence that it will be applied without breaking the whole system.

Before CI, changes to the software were often made in large batches and often required a long validation process. It can be months before developers see their changes in production, receive feedback loops, and correct any bugs. In short, the CI process aims to shorten the process from idea to change.

The general workflow typically involves the following steps:

1. The first engineer takes a current copy of the code base and works on their change
2. The first engineer submits the change to the repository
3. The repository can notify the necessary parties of a change in the repository to a group of engineers who can review the change. They can either approve or reject the change
4. The continuous-integration system can continuously pull the repository for changes, or the repository can send a notification to the CI system when changes happen. Either way, the CI system will pull the latest version of the code

5. The CI system will run automated tests to try to catch any breakage
6. If there is no fault found, the CI system can choose to merge the change into the main code and optionally deploy to the production system

This is a generalized list of steps. The process can be different for each organization; for example, automated tests can be run as soon as the delta code is checked in instead of after code review. Sometimes, the organization might choose to have a human engineer involved for sanity checks in between the steps.

In the next section, we will illustrate the instructions to install Jenkins on an Ubuntu 16.04 system.

Installing Jenkins

For the examples we will use in this chapter, we can install Jenkins on the management host or a separate machine. My personal preference is to install it on a separate virtual machine. The virtual machine will have a similar network set up as the management host up to this point, with one interface for the internet connection and another interface for VMNet 2 connection to the VIRL management network.

The Jenkins image and installation instruction per operating system can be found at https://jenkins.io/download/. The following is the instructions I used for installing Jenkins on the Ubuntu 16.04 host:

```
$ wget -q -O - https://pkg.jenkins.io/debian-stable/jenkins.io.key | sudo apt-key add -

# added Jenkins to /etc/apt/sources.list
$ cat /etc/apt/sources.list | grep jenkins
deb https://pkg.jenkins.io/debian-stable binary/

# install Java8
$ sudo add-apt-repository ppa:webupd8team/java
$ sudo apt update; sudo apt install oracle-java8-installer

$ sudo apt-get update
$ sudo apt-get install jenkins

# Start Jenkins
$ /etc/init.d/jenkins start
```

Chapter 12

 At the time of writing, we have to install Java separately because Jenkins does not work with Java 9; see `https://issues.jenkins-ci.org/browse/JENKINS-40689` for more details. Hopefully, by the time you read this, the issue is resolved.

Once Jenkins is installed, we can point the browser to the IP at port `8080` to continue the process:

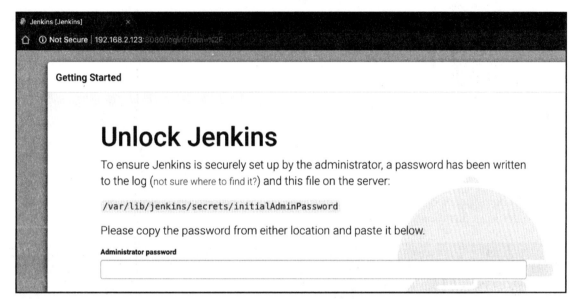

Unlock Jenkins screen

[385]

Continuous Integration with Jenkins

As stated on the screen, get the admin password from `/var/lib/jenkins/secrets/initialAdminPassword` and paste the output in the screen. For the time being, we will choose the **Install suggested plugins** option:

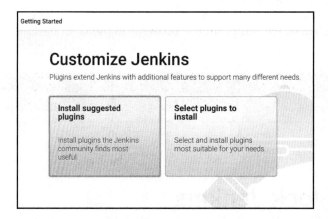

Install suggested plugins

You will be redirected to create the admin user; once created, Jenkins will be ready. If you see the Jenkins dashboard, the installation was successful:

Jenkins dashboard

Chapter 12

We are now ready to use Jenkins to schedule our first job.

Jenkins example

In this section, we will take a look at a few Jenkins examples and how they tie into the various technologies we have covered in this book. The reason Jenkins is one of the last chapters of this book is because it will leverage many of the other tools, such as our Python script, Ansible, Git, and GitHub. Feel free to refer back to `Chapters 11`, *Working with Git*, if needed.

 In the examples, we will use the Jenkins master to execute our jobs. In production, it is recommended to add Jenkins nodes to handle the execution of jobs.

For our lab, we will use a simple two-node topology with IOSv devices:

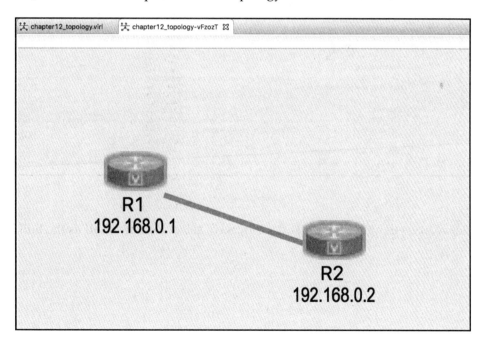

Chapter 12 lab topology

Let's build our first job.

First job for the Python script

For our first job, let's use the Parmiko script that we built in Chapter 2, *Low-Level Network Device Interactions*, chapter2_3.py. If you recall, this is a script that uses Paramiko to ssh to the remote devices and grabs the show run and show version output of the devices:

```
$ ls
chapter12_1.py
$ python3 /home/echou/Chapter12/chapter12_1.py
...
$ ls
chapter12_1.py iosv-1_output.txt iosv-2_output.txt
```

We will use the create new job link to create the job and pick the **Freestyle project** option:

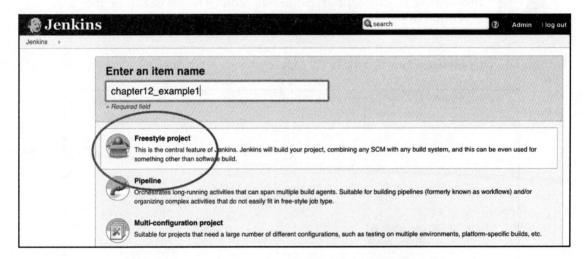

Example 1 freestyle project

We will leave everything as default and unchecked; select **Execute shell** as the build option:

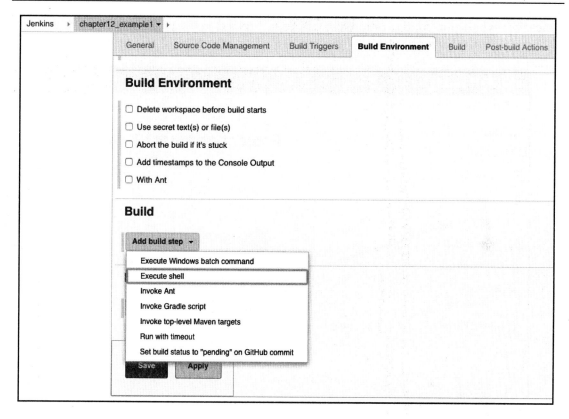

Example 1 build step

When the prompt appears, we will enter in the exact commands we use in the shell:

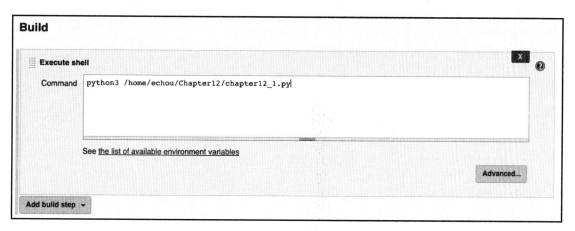

Example 1 shell command

Once we save the job configuration, we will be redirected to the project dashboard. We can choose the **Build Now** option, and the job will appear under **Build History**:

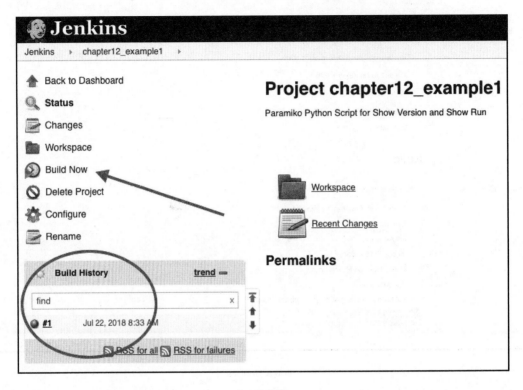

Example 1 build

You can check the status of the build by clicking on it and choosing the **Console Output** on the left panel:

Chapter 12

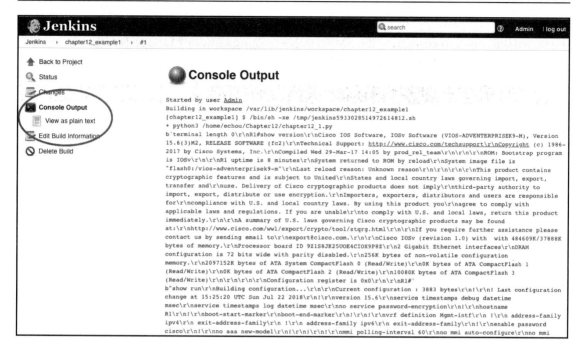

Example 1 console output

As an optional step, we can schedule this job at a regular interval, much like cron would do for us. The job can be scheduled under **Build Triggers**, choose to **Build Periodically** and entered the cron-like schedule. In this example, the script will run daily at 02:00 and 22:00:

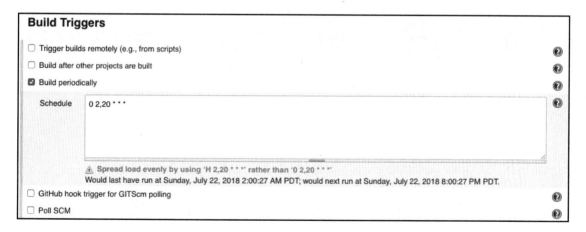

Example 1 build trigger

[391]

Continuous Integration with Jenkins

We can also configure the SMTP server on Jenkins to allow notification of the build results. First, we will need to configure the SMTP server settings under **Manage Jenkins** | **Configure Systems** from the main menu:

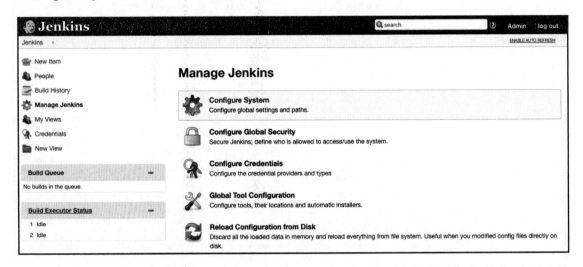

Example 1 configure system

We will see the SMTP server settings toward the bottom of the page. Click on the **Advanced settings** to configure the **SMTP server** settings as well as to send out a test email:

Chapter 12

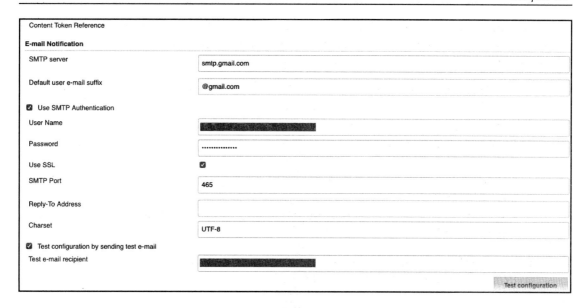

Example 1 configure SMTP

We will be able to configure email notifications as part of the post-build actions for our job:

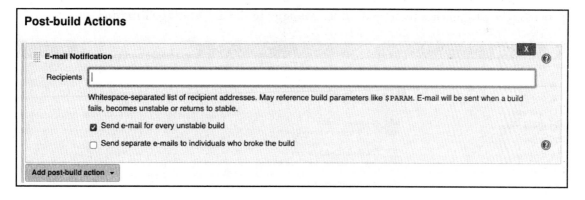

Example 1 email notification

Congratulations! We have just used Jenkins to create our first job. Functionally, this has not done anything more than what we could have achieved with our management host. However, there are several advantages of using Jenkins:

- We can utilize Jenkins' various database-authentication integrations, such as LDAP, to allow existing users to execute our script.
- We can use Jenkins' role-based authorization to limit users. For example, some users can only execute jobs without modification access while others can have full administrative access.
- Jenkins provides a web-based graphical interface that allows users to access, the scripts easily.
- We can use the Jenkins email and logging services to centralize our jobs and be notified of the results.

Jenkins is a great tool by itself. Just like Python, it has a big third-party plugin ecosystem that can be used to expand its features and functionalities.

Jenkins plugins

We will install a simple schedule plugin as an example illustrating the plugin-installation process. The plugins are managed under **Manage Jenkins** | **Manage Plugins**:

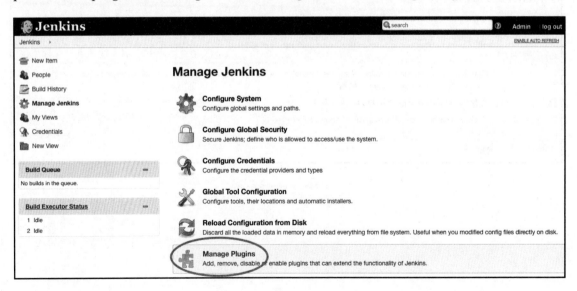

Jenkins plugin

We can use the search function to look for the **Schedule Build** plugin under the available tab:

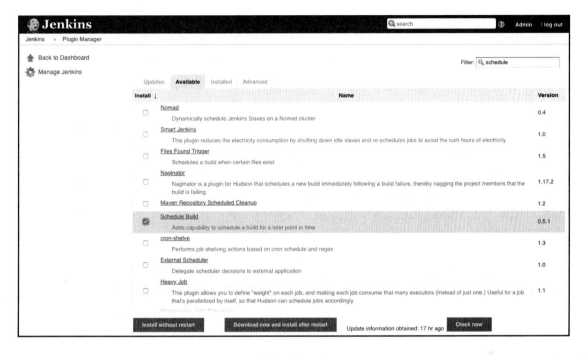

Jenkins plugin search

From there, we will just click on **Install without restart**, and we will be able to check the installation progress on the following page:

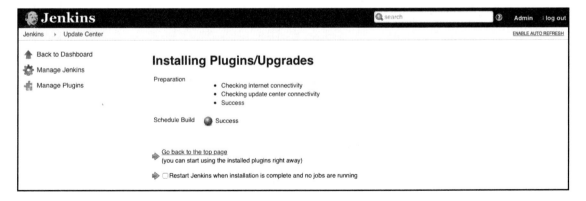

Jenkins plugin installation

Continuous Integration with Jenkins

After the installation is completed, we will be able to see a new icon that allows us to schedule jobs more intuitively:

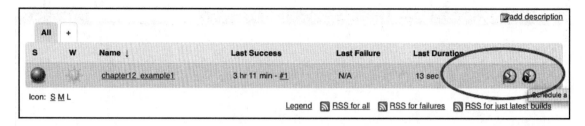

Jenkins plugin result

It is one of the strengths of a popular open source project to have the ability to grow over time. For Jenkins, the plugins provide a way to customize the tool for different customer needs. In the coming section, we will look at how to integrate version control and the approval process into our workflow.

Network continuous integration example

In this section, let's integrate our GitHub repository with Jenkins. By integrating the GitHub repository, we can take advantage of the GitHub code review and collaboration tools.

First, we will create a new GitHub repository, I will call this repository `chapter12_example2`. We can clone this repository locally and add the files we wanted to the repository. In this case, I am adding an Ansible playbook that copies the output of the `show version` command to a file:

```
$ cat chapter12_playbook.yml
---
- name: show version
  hosts: "ios-devices"
  gather_facts: false
  connection: local

  vars:
    cli:
      host: "{{ ansible_host }}"
      username: "{{ ansible_user }}"
      password: "{{ ansible_password }}"

  tasks:
    - name: show version
      ios_command:
```

[396]

```
            commands: show version
            provider: "{{ cli }}"

      register: output

    - name: show output
      debug:
          var: output.stdout

    - name: copy output to file
      copy: content="{{ output }}" dest=./output/{{ inventory_hostname }}.txt
```

By now, we should be pretty familiar with running an Ansible playbook. I will skip the output of `host_vars` and the inventory file. However, the most important thing is to verify that it runs on the local machine before committing to the GitHub repository:

```
$ ansible-playbook -i hosts chapter12_playbook.yml

PLAY [show version]
*************************************************************

TASK [show version]
*************************************************************
ok: [iosv-1]
ok: [iosv-2]
...
TASK [copy output to file]
*********************************************************
changed: [iosv-1]
changed: [iosv-2]

PLAY RECAP
******************************************************************
iosv-1 : ok=3 changed=1 unreachable=0 failed=0
iosv-2 : ok=3 changed=1 unreachable=0 failed=0
```

We can now push the playbook and associated files to our GitHub repository:

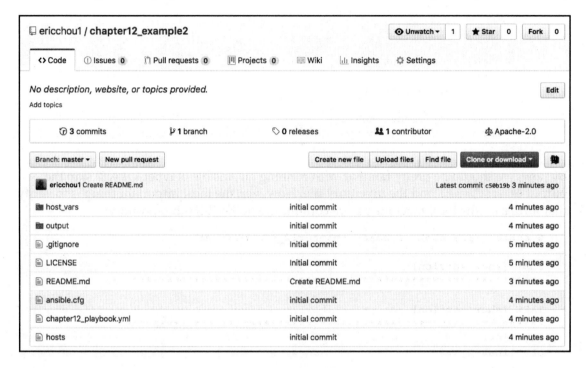

Example 2 GitHub repository

Let's log back into the Jenkins host to install `git` and Ansible:

```
$ sudo apt-get install git
$ sudo apt-get install software-properties-common
$ sudo apt-get update
$ sudo apt-get install ansible
```

Chapter 12

Some of the tools can be installed under **Global Tool Configuration**; Git is one of them. However, since we are installing Ansible, we can install Git in the same Command Prompt:

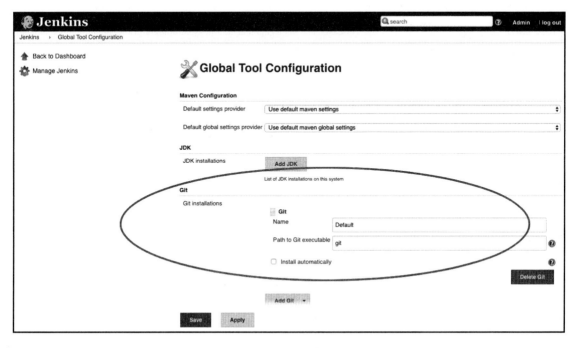

Global tools configuration

We can create a new freestyle project named `chapter12_example2`. Under the source-code management, we will specify the GitHub repository as the source:

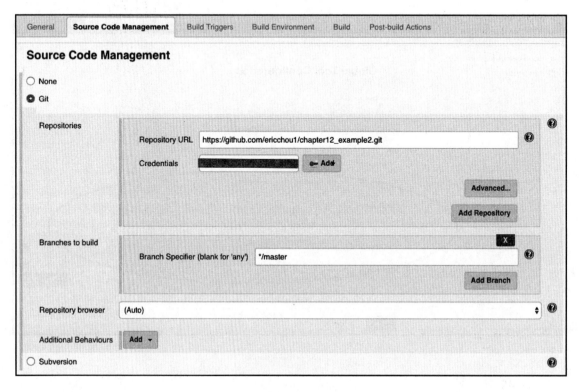

Example 2 source-code management

Chapter 12

Before we move on to the next step, let's save the project and run a build. In the build console output, we should be able to see the repository being cloned and the index value match what we see on GitHub:

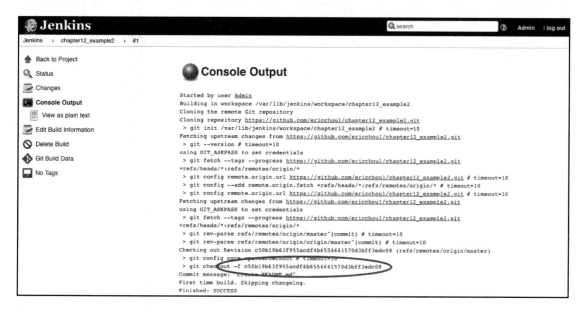

Example 2 console output 1

We can now add the Ansible playbook command in the build section:

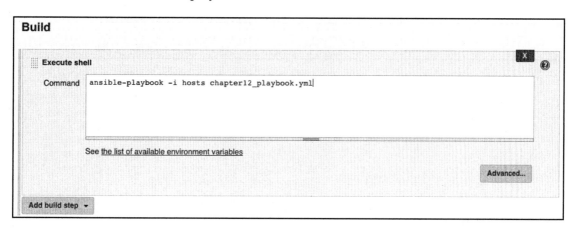

Example 2 build shell

[401]

If we run the build again, we can see from the console output that Jenkins will fetch the code from GitHub before executing the Ansible playbook:

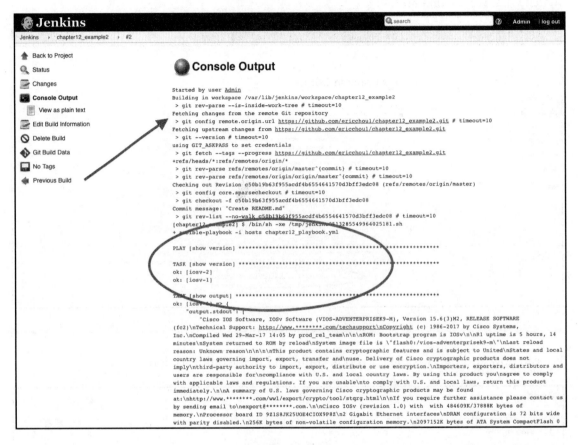

Example 2 build console output 2

One of the benefits of integrating GitHub with Jenkins is that we can see all the Git information on the same screen:

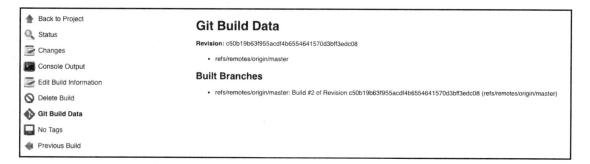

Example 2 Git build data

The results of the project, such as the output of the Ansible playbook, can be seen in the `workspace` folder:

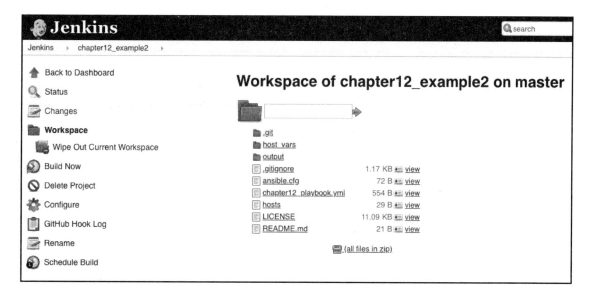

Example 2 workspace

Continuous Integration with Jenkins

At this point, we can follow the same step as before to use periodic build as the build trigger. If the Jenkins host is publicly accessible, we can also use GitHub's Jenkins plugin to notify Jenkins as a trigger for the build. This is a two-step process, the first step is to enable the plugin on your GitHub repository:

Example 2 GitHub Jenkins service

The second step is to specify the GitHub hook trigger as the **Build Triggers** for our project:

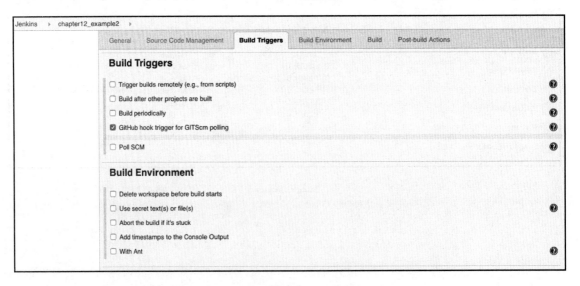

Example 2 Jenkins build trigger

Having the GitHub repository as the source allows for a brand new set of possibilities of treating infrastructure as code. We can now use GitHub's tool of a fork, pull requests, issue tracking, and project management to work together efficiently. Once the code is ready, Jenkins can automatically pull the code down and execute it on our behalf.

 You will notice we did not mention anything about automated testing. We will go over testing in Chapter 13, *Test-Driven Development for Networks*.

Jenkins is a full-featured system that can become complex. We have just scratched the surface of it with the two examples presented in this chapter. The Jenkins pipeline, environmental setup, multibranch pipeline, and so on, are all useful features that can accommodate the most complex automation projects. Hopefully, this chapter will serve as an interesting introduction for you to further explore the Jenkins tool.

Jenkins with Python

Jenkins provides a full set of REST APIs for its functionalities: https://wiki.jenkins.io/display/JENKINS/Remote+access+API. There are also a number of Python wrappers that make the interaction even easier. Let's take a look at the Python-Jenkins package:

```
$ sudo pip3 install python-jenkins
$ python3
>>> import jenkins
>>> server = jenkins.Jenkins('http://192.168.2.123:8080', username='<user>', password='<pass>')
>>> user = server.get_whoami()
>>> version = server.get_version()
>>> print('Hello %s from Jenkins %s' % (user['fullName'], version))
Hello Admin from Jenkins 2.121.2
```

We can work with the management of the server, such as plugins:

```
>>> plugin = server.get_plugins_info()
>>> plugin
[{'supportsDynamicLoad': 'MAYBE', 'downgradable': False,
'requiredCoreVersion': '1.642.3', 'enabled': True, 'bundled': False,
'shortName': 'pipeline-stage-view', 'url':
'https://wiki.jenkins-ci.org/display/JENKINS/Pipeline+Stage+View+Plugin',
'pinned': False, 'version': 2.10, 'hasUpdate': False, 'deleted': False,
'longName': 'Pipeline: Stage View Plugin', 'active': True, 'backupVersion':
None, 'dependencies': [{'shortName': 'pipeline-rest-api', 'version':
'2.10', 'optional': False}, {'shortName': 'workflow-job', 'version': '2.0',
```

We can also manage the Jenkins jobs:

```
>>> job = server.get_job_config('chapter12_example1')
>>> import pprint
>>> pprint.pprint(job)
("<?xml version='1.1' encoding='UTF-8'?>\n"
 '<project>\n'
 '  <actions/>\n'
 '  <description>Paramiko Python Script for Show Version and Show '
 'Run</description>\n'
 '  <keepDependencies>false</keepDependencies>\n'
 '  <properties>\n'
 '    <jenkins.model.BuildDiscarderProperty>\n'
 '      <strategy class="hudson.tasks.LogRotator">\n'
 '        <daysToKeep>10</daysToKeep>\n'
 '        <numToKeep>5</numToKeep>\n'
 '        <artifactDaysToKeep>-1</artifactDaysToKeep>\n'
 '        <artifactNumToKeep>-1</artifactNumToKeep>\n'
 '      </strategy>\n'
 '    </jenkins.model.BuildDiscarderProperty>\n'
 '  </properties>\n'
 '  <scm class="hudson.scm.NullSCM"/>\n'
 '  <canRoam>true</canRoam>\n'
 '  <disabled>false</disabled>\n'
 '  '
 '<blockBuildWhenDownstreamBuilding>false</blockBuildWhenDownstreamBuilding>\n'
 '  '
 '<blockBuildWhenUpstreamBuilding>false</blockBuildWhenUpstreamBuilding>\n'
 '  <triggers>\n'
 '    <hudson.triggers.TimerTrigger>\n'
 '      <spec>0 2,20 * * *</spec>\n'
 '    </hudson.triggers.TimerTrigger>\n'
 '  </triggers>\n'
 '  <concurrentBuild>false</concurrentBuild>\n'
 '  <builders>\n'
 '    <hudson.tasks.Shell>\n'
 '      <command>python3 /home/echou/Chapter12/chapter12_1.py</command>\n'
 '    </hudson.tasks.Shell>\n'
 '  </builders>\n'
 '  <publishers/>\n'
 '  <buildWrappers/>\n'
 '</project>')
>>>
```

Using Python-Jenkins allows us to have a way to interact with Jenkins in a programmatic way.

Continuous integration for Networking

Continuous integration has been adopted in the software-development world for a while, but it is relatively new to network engineering. We are admittedly a bit behind in terms of using continuous integration in our network infrastructure. It is no doubt a bit of a challenge to think of our network in terms of code when we are still struggling to figure out how to stop using the CLI to manage our devices.

There are a number of good examples of using Jenkins for network automation. One is by Tim Fairweather and Shea Stewart at AnsibleFest 2017 network track: `https://www.ansible.com/ansible-for-networks-beyond-static-config-templates`. Another use case was shared by Carlos Vicente from Dyn at NANOG 63: `https://www.nanog.org/sites/default/files/monday_general_autobuild_vicente_63.28.pdf`.

Even though continuous integration might be an advanced topic for network engineers who are just beginning to learn coding and the toolsets, in my opinion, it is worth the effort to start learning and using continuous integration in production today. Even at the basic level, the experience will trigger more innovative ways for network automation that will no doubt help the industry move forward.

Summary

In this chapter, we examined the traditional change-management process and why it is not a good fit for today's rapidly changing environment. The network needs to evolve with the business to become more agile and adapt to change quickly and reliably.

We looked at the concept of continuous integration, in particular the open source Jenkins system. Jenkins is a full-featured, expandable, continuous-integration system that is widely used in software development. We installed and used Jenkins to execute our Python script based on `Paramiko` in a periodic interval with email notifications. We also saw how we can install plugins for Jenkins to expand its features.

We looked at how we can use Jenkins to integrate with our GitHub repository and trigger builds based on code-checking. By integrating Jenkins with GitHub, we can utilize the GitHub process of collaboration.

In `Chapter 13`, *Test-Driven Development for Networks*, we will look at test-driven development with Python.

13
Test-Driven Development for Networks

The idea of **Test-Driven Development** (**TDD**) has been around for a while. American software engineer Kent Beck, among others, is typically credited with bringing and leading the TDD movement along with agile software development. Agile software development requires very short build-test-deploy development cycles; all of the software requirements are turned into test cases. These test cases are usually written before the code is written, and the software code is only accepted when the test passes.

The same idea can be drawn in parallel with network engineering. When we face the challenge of designing a modern network, we can break the process down into the following steps:

- We start with the overall requirement for the new network. Why do we need to design a new or part of a new network? Maybe it is for new server hardware, a new storage network, or a new micro-service software architecture.
- The new requirements are broken down into smaller, more specific requirements. This can be looking at a new switch platform, a more efficient routing protocol, or a new network topology (for example, fat-tree). Each of the smaller requirements can be broken down into the categories of must-have and optional.
- We draw out the test plan and evaluate it against the potential candidates for solutions.
- The test plan will work in reverse order; we will start by testing the features, then integrate the new feature into a bigger topology. Finally, we will try to run our test as close to a production environment as possible.

The point is, even if we don't realize it, we might already be adopting a test-driven development methodology in network engineering. This was part of my revelation when I was studying the TDD mindset. We are already implicitly following this best practice without formalizing the method.

By gradually moving parts of the network as code, we can use TDD for the network even more. If our network topology is described in a hierarchical format in XML or JSON, each of the components can be correctly mapped and expressed in the desired state. This is the desired state that we can write test cases against. For example, if our desired state calls for a full mesh of switches, we can always write a test case to check against our production devices for the number of BGP neighbors it has.

Test-driven development overview

The sequence of TDD is loosely based on the following six steps:

1. Write a test with the result in mind
2. Run all tests and see whether the new test fails
3. Write the code
4. Run the test again
5. Make necessary changes if the test fails
6. Repeat

I just follow the guidelines loosely. The TDD process calls for writing the test cases before writing any code, or in our instance, before any components of the network are built. As a matter of personal preference, I always like to see a working version of the working network or code before writing test cases. It gives me a higher level of confidence. I also jump around the levels of testing; sometimes I test a small portion of the network; other times I conduct a system-level end-to-end test, such as a ping or traceroute test.

The point is, I do not believe there is a one-size-fits-all approach when it comes to testing. It depends on personal preference and the scope of the project. This is true for most of the engineers I have worked with. It is a good idea to keep the framework in mind, so we have a working blueprint to follow, but you are the best judge of your style of problem-solving.

Test definitions

Let's look at some of the terms commonly used in TDD:

- **Unit test**: Checks a small piece of code. This is a test that is run against a single function or class
- **Integration test**: Checks multiple components of a code base; multiple units are combined and tested as a group. This can be a test that checks against a Python module or multiple modules
- **System test**: Checks from end to end. This is a test that runs as close to what an end user would see
- **Functional test**: Checks against a single function
- **Test coverage**: A term defined as the determination of whether our test cases cover the application code. This is typically done by examining how much code is exercised when we run the test cases
- **Test fixtures**: A fixed state that forms a baseline for running our tests. The purpose of a test fixture is to ensure there is a well-known and fixed environment in which tests are run, so they are repeatable
- **Setup and teardown**: All the prerequisite steps are added in the setup and cleaned up in the teardown

The terms might seem very software-development-centric, and some might not be relevant to network engineering. Keep in mind that the terms are a way for us to communicate a concept or step we will be using these terms in the rest of this chapter. As we use the terms more in the network engineering context, they might become clearer. Let's dive into treating network topology as code.

Topology as code

Before we declare that the network is too complex, it is impossible to summarize it into code! Let's keep an open mind. Would it help if I tell you we have been using code to describe our topology in this book already?

If you take a look at any of the VIRL topology graphs that we have been using in this book, they are simply XML files that include a description of the relationship between nodes.

In this chapter, we will use the following topology for our lab:

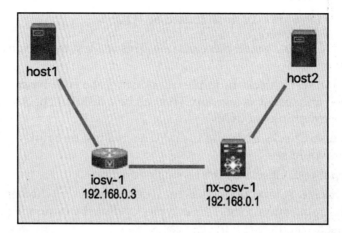

If we open up the topology file, chapter13_topology.virl, with a text editor, we will see that the file is an XML file describing the node and the relationship between the nodes. The top root level is the <topology> node with child nodes of <node>. Each of the child nodes consists of various extensions and entries. The device configurations are embedded in the file as well:

```
<?xml version="1.0" encoding="UTF-8" standalone="yes"?>
<topology xmlns="http://www.cisco.com/VIRL"
xmlns:xsi="http://www.w3.org/2001/XMLSchema-instance" schemaVersion="0.95"
xsi:schemaLocation="http://www.cisco.com/VIRL
https://raw.github.com/CiscoVIRL/schema/v0.95/virl.xsd">
    <extensions>
        <entry key="management_network" type="String">flat</entry>
    </extensions>
    <node name="iosv-1" type="SIMPLE" subtype="IOSv" location="182,162"
ipv4="192.168.0.3">
        <extensions>
            <entry key="static_ip" type="String">172.16.1.20</entry>
            <entry key="config" type="string">! IOS Config generated on
2018-07-24 00:23
! by autonetkit_0.24.0
!
hostname iosv-1
boot-start-marker
boot-end-marker
!
...
    </node>
    <node name="nx-osv-1" type="SIMPLE" subtype="NX-OSv" location="281,161"
```

```
    ipv4="192.168.0.1">
        <extensions>
            <entry key="static_ip" type="String">172.16.1.21</entry>
            <entry key="config" type="string">! NX-OSv Config generated on
2018-07-24 00:23
! by autonetkit_0.24.0
!
version 6.2(1)
license grace-period
!
hostname nx-osv-1

...
    <node name="host2" type="SIMPLE" subtype="server" location="347,66">
        <extensions>
            <entry key="static_ip" type="String">172.16.1.23</entry>
            <entry key="config" type="string">#cloud-config
bootcmd:
- ln -s -t /etc/rc.d /etc/rc.local
hostname: host2
manage_etc_hosts: true
runcmd:
- start ttyS0
- systemctl start getty@ttyS0.service
- systemctl start rc-local
      <annotations/>
      <connection dst="/virl:topology/virl:node[1]/virl:interface[1]"
src="/virl:topology/virl:node[3]/virl:interface[1]"/>
      <connection dst="/virl:topology/virl:node[2]/virl:interface[1]"
src="/virl:topology/virl:node[1]/virl:interface[2]"/>
      <connection dst="/virl:topology/virl:node[4]/virl:interface[1]"
src="/virl:topology/virl:node[2]/virl:interface[2]"/>
</topology>
```

By expressing the network as code, we can declare a source of truth for our network. We can write test code to compare the actual production value against this blueprint. We will use this topology file as the base, and compare the production network value against it. But first, we will need to grab the values we want from the XML file. In `chapter13_1_xml.py`, we will use `ElementTree` to parse the `virl` topology file and construct a dictionary consisting of the information of our devices:

```
#!/usr/env/bin python3

import xml.etree.ElementTree as ET
import pprint

with open('chapter13_topology.virl', 'rt') as f:
```

```
        tree = ET.parse(f)

devices = {}

for node in tree.findall('./{http://www.cisco.com/VIRL}node'):
    name = node.attrib.get('name')
    devices[name] = {}
    for attr_name, attr_value in sorted(node.attrib.items()):
        devices[name][attr_name] = attr_value

# Custom attributes
devices['iosv-1']['os'] = '15.6(3)M2'
devices['nx-osv-1']['os'] = '7.3(0)D1(1)'
devices['host1']['os'] = '16.04'
devices['host2']['os'] = '16.04'

pprint.pprint(devices)
```

The result is a Python dictionary that consists of the devices according to our topology file. We can also add customary items to the dictionary:

```
$ python3 chapter13_1_xml.py
{'host1': {'location': '117,58',
           'name': 'host1',
           'os': '16.04',
           'subtype': 'server',
           'type': 'SIMPLE'},
 'host2': {'location': '347,66',
           'name': 'host2',
           'os': '16.04',
           'subtype': 'server',
           'type': 'SIMPLE'},
 'iosv-1': {'ipv4': '192.168.0.3',
            'location': '182,162',
            'name': 'iosv-1',
            'os': '15.6(3)M2',
            'subtype': 'IOSv',
            'type': 'SIMPLE'},
 'nx-osv-1': {'ipv4': '192.168.0.1',
              'location': '281,161',
              'name': 'nx-osv-1',
              'os': '7.3(0)D1(1)',
              'subtype': 'NX-OSv',
              'type': 'SIMPLE'}}
```

We can use our example from Chapter 3, *APIs and Intent-Driven Networking*, cisco_nxapi_2.py, to retrieve the NX-OSv version. When we combine the two files, we can compare the value we received from our topology file as well as the production device information. We can use Python's built-in unittest module to write test cases.

We will discuss the unittest module later. Feel free to skip ahead and come back to this example if you'd like.

Here is the relevant unittest portion in chapter13_2_validation.py:

```
import unittest

# Unittest Test case
class TestNXOSVersion(unittest.TestCase):
    def test_version(self):
        self.assertEqual(nxos_version, devices['nx-osv-1']['os'])

if __name__ == '__main__':
    unittest.main()
```

When we run the validation test, we can see that the test passes because the software version in production matches what we expected:

```
$ python3 chapter13_2_validation.py
.
----------------------------------------------------------------------
Ran 1 test in 0.000s

OK
```

If we manually change the expected NX-OSv version value to introduce a failure case, we will see the following failed output:

```
$ python3 chapter13_3_test_fail.py
F
======================================================================
FAIL: test_version (__main__.TestNXOSVersion)
----------------------------------------------------------------------
Traceback (most recent call last):
  File "chapter13_3_test_fail.py", line 50, in test_version
    self.assertEqual(nxos_version, devices['nx-osv-1']['os'])
AssertionError: '7.3(0)D1(1)' != '7.4(0)D1(1)'
- 7.3(0)D1(1)
?   ^
```

```
+ 7.4(0)D1(1)
?    ^
```

```
----------------------------------------------------------------
Ran 1 test in 0.004s

FAILED (failures=1)
```

We can see that the test case result was returned as failed; the reason for failure was the version mismatch between the two values.

Python's unittest module

In the previous example, we saw how we could use the `assertEqual()` method to compare the two values to return either `True` or `False`. Here is an example of the built-in `unittest` module to compare two values:

```
$ cat chapter13_4_unittest.py
#!/usr/bin/env python3

import unittest

class SimpleTest(unittest.TestCase):
    def test(self):
        one = 'a'
        two = 'a'
        self.assertEqual(one, two)
```

Using the `python3` command-line interface, the `unittest` module can automatically discover the test cases in the script:

```
$ python3 -m unittest chapter13_4_unittest.py
.
----------------------------------------------------------------
Ran 1 test in 0.000s

OK
```

Besides comparing two values, here are more examples of testing if the expected value is `True` or `False`. We can also generate custom failure messages when a failure occurs:

```
$ cat chapter13_5_more_unittest.py
#!/usr/bin/env python3
# Examples from https://pymotw.com/3/unittest/index.html#module-unittest

import unittest

class Output(unittest.TestCase):
    def testPass(self):
        return

    def testFail(self):
        self.assertFalse(True, 'this is a failed message')

    def testError(self):
        raise RuntimeError('Test error!')

    def testAssesrtTrue(self):
        self.assertTrue(True)

    def testAssertFalse(self):
        self.assertFalse(False)
```

We can use `-v` for the option to display a more detailed output:

```
$ python3 -m unittest -v chapter13_5_more_unittest.py
testAssertFalse (chapter13_5_more_unittest.Output) ... ok
testAssesrtTrue (chapter13_5_more_unittest.Output) ... ok
testError (chapter13_5_more_unittest.Output) ... ERROR
testFail (chapter13_5_more_unittest.Output) ... FAIL
testPass (chapter13_5_more_unittest.Output) ... ok

======================================================================
ERROR: testError (chapter13_5_more_unittest.Output)
----------------------------------------------------------------------
Traceback (most recent call last):
  File "/home/echou/Master_Python_Networking_second_edition/Chapter13/chapter13_5_more_unittest.py", line 14, in testError
    raise RuntimeError('Test error!')
RuntimeError: Test error!

======================================================================
FAIL: testFail (chapter13_5_more_unittest.Output)
----------------------------------------------------------------------
Traceback (most recent call last):
```

```
        File
"/home/echou/Master_Python_Networking_second_edition/Chapter13/chapter13_5_
more_unittest.py", line 11, in testFail
        self.assertFalse(True, 'this is a failed message')
AssertionError: True is not false : this is a failed message

----------------------------------------------------------------------
Ran 5 tests in 0.001s

FAILED (failures=1, errors=1)
```

Starting from Python 3.3, the `unittest` module includes the `module` object library by default (https://docs.python.org/3/library/unittest.mock.html). This is a very useful module to make a fake HTTP API call to a remote resource without actually making the call. For example, we have seen the example of using NX-API to retrieve the NX-OS version number. What if we want to run our test, but we do not have an NX-OS device available? We can use the `unittest` mock object.

In `chapter13_5_more_unittest_mocks.py`, we created a simple class with a method to make HTTP API calls and expect a JSON response:

```
# Our class making API Call using requests
class MyClass:
    def fetch_json(self, url):
        response = requests.get(url)
        return response.json()
```

We also created a function that mocks two URL calls:

```
# This method will be used by the mock to replace requests.get
def mocked_requests_get(*args, **kwargs):
    class MockResponse:
        def __init__(self, json_data, status_code):
            self.json_data = json_data
            self.status_code = status_code

        def json(self):
            return self.json_data

    if args[0] == 'http://url-1.com/test.json':
        return MockResponse({"key1": "value1"}, 200)
    elif args[0] == 'http://url-2.com/test.json':
        return MockResponse({"key2": "value2"}, 200)

    return MockResponse(None, 404)
```

Finally, we make the API call to the two URLs in our test case. However, we are using the `mock.patch` decorator to intercept the API calls:

```
# Our test case class
class MyClassTestCase(unittest.TestCase):
    # We patch 'requests.get' with our own method. The mock object is
    passed in to our test case method.
    @mock.patch('requests.get', side_effect=mocked_requests_get)
    def test_fetch(self, mock_get):
        # Assert requests.get calls
        my_class = MyClass()
        # call to url-1
        json_data = my_class.fetch_json('http://url-1.com/test.json')
        self.assertEqual(json_data, {"key1": "value1"})
        # call to url-2
        json_data = my_class.fetch_json('http://url-2.com/test.json')
        self.assertEqual(json_data, {"key2": "value2"})
        # call to url-3 that we did not mock
        json_data = my_class.fetch_json('http://url-3.com/test.json')
        self.assertIsNone(json_data)


if __name__ == '__main__':
    unittest.main()
```

When we run the test, we will see that the test passes without needing to make an actual API call to the remote endpoint:

```
$ python3 -m unittest -v chapter13_5_more_unittest_mocks.py
test_fetch (chapter13_5_more_unittest_mocks.MyClassTestCase) ... ok

----------------------------------------------------------------------
Ran 1 test in 0.001s

OK
```

For more information on the `unittest` module, Doug Hellmann's Python module of the week (https://pymotw.com/3/unittest/index.html#module-unittest) is an excellent source of short and precise examples on the `unittest` module. As always, the Python documentation is a good source of information as well: https://docs.python.org/3/library/unittest.html.

More on Python testing

In addition to the built-in library of `unittest`, there are lots of other Python testing frameworks in the community. Pytest is another robust Python testing framework that is worth a look. `pytest` can be used for all types and levels of software testing. It can be used by developers, QA engineers, individuals practicing Test-Driven Development, and open source projects. Many of the large-scale open source projects have switched from `unittest` or `nose` to `pytest`, including Mozilla and Dropbox. The main attractive features of `pytest` were a third-party plugin model, a simple fixture model, and assert rewriting.

> If you want to learn more about the `pytest` framework, I would highly recommend *Python Testing with PyTest* by Brian Okken (ISBN 978-1-68050-240-4). Another great source is the `pytest` documentation: https://docs.pytest.org/en/latest/.

`pytest` is command-line-driven; it can find the tests we have written automatically and run them:

```
$ sudo pip install pytest
$ sudo pip3 install pytest
$ python3
Python 3.5.2 (default, Nov 23 2017, 16:37:01)
[GCC 5.4.0 20160609] on linux
Type "help", "copyright", "credits" or "license" for more information.
>>> import pytest
>>> pytest.__version__
'3.6.3'
```

Let's look at some examples using `pytest`.

pytest examples

The first `pytest` example will be a simple assert for two values:

```
$ cat chapter13_6_pytest_1.py
#!/usr/bin/env python3

def test_passing():
    assert(1, 2, 3) == (1, 2, 3)

def test_failing():
    assert(1, 2, 3) == (3, 2, 1)
```

When you run with the -v option, pytest will give us a pretty robust answer for the failure reason:

```
$ pytest -v chapter13_6_pytest_1.py
============================ test session starts ============================
platform linux -- Python 3.5.2, pytest-3.6.3, py-1.5.4, pluggy-0.6.0 -- /usr/bin/python3
cachedir: .pytest_cache
rootdir: /home/echou/Master_Python_Networking_second_edition/Chapter13, inifile:
collected 2 items

chapter13_6_pytest_1.py::test_passing PASSED                           [ 50%]
chapter13_6_pytest_1.py::test_failing FAILED                           [100%]

================================== FAILURES ==================================
_____ test_failing _____

    def test_failing():
>       assert(1, 2, 3) == (3, 2, 1)
E       assert (1, 2, 3) == (3, 2, 1)
E         At index 0 diff: 1 != 3
E         Full diff:
E         - (1, 2, 3)
E         ?    ^  ^
E         + (3, 2, 1)
E         ?    ^  ^

chapter13_6_pytest_1.py:7: AssertionError
===================== 1 failed, 1 passed in 0.03 seconds =====================
```

In the second example, we will create a `router` object. The `router` object will be initiated with some values in `None` and some values with default values. We will use `pytest` to test one instance with the default and one instance without:

```
$ cat chapter13_7_pytest_2.py
#!/usr/bin/env python3

class router(object):
    def __init__(self, hostname=None, os=None, device_type='cisco_ios'):
        self.hostname = hostname
        self.os = os
        self.device_type = device_type
        self.interfaces = 24
```

```python
def test_defaults():
    r1 = router()
    assert r1.hostname == None
    assert r1.os == None
    assert r1.device_type == 'cisco_ios'
    assert r1.interfaces == 24

def test_non_defaults():
    r2 = router(hostname='lax-r2', os='nxos', device_type='cisco_nxos')
    assert r2.hostname == 'lax-r2'
    assert r2.os == 'nxos'
    assert r2.device_type == 'cisco_nxos'
    assert r2.interfaces == 24
```

When we run the test, we will see whether the instance was accurately applied with the default values:

```
$ pytest chapter13_7_pytest_2.py
=============================== test session starts ===============================
platform linux -- Python 3.5.2, pytest-3.6.3, py-1.5.4, pluggy-0.6.0
rootdir: /home/echou/Master_Python_Networking_second_edition/Chapter13,
inifile:
collected 2 items

chapter13_7_pytest_2.py ..                                                 [100%]

============================ 2 passed in 0.04 seconds =============================
```

If we were to replace the previous `unittest` example with `pytest`, in `chapter13_8_pytest_3.py` we will have a simple test case:

```python
# pytest test case
def test_version():
    assert devices['nx-osv-1']['os'] == nxos_version
```

Then we run the test with the `pytest` command line:

```
$ pytest chapter13_8_pytest_3.py
=============================== test session starts ===============================
platform linux -- Python 3.5.2, pytest-3.6.3, py-1.5.4, pluggy-0.6.0
rootdir: /home/echou/Master_Python_Networking_second_edition/Chapter13,
inifile:
collected 1 item

chapter13_8_pytest_3.py .                                                  [100%]
```

```
============================ 1 passed in 0.19 seconds
============================
```

If we are writing tests for ourselves, we are free to choose any modules. Between `unittest` and `pytest`, I find `pytest` a more intuitive tool to use. However, since `unittest` is included in the standard library, many teams might have a preference for using the `unittest` module for their testing.

Writing tests for networking

So far, we have been mostly writing tests for our Python code. We have used both the `unittest` and `pytest` libraries to assert `True/False` and `equal/Non-equal` values. We were also able to write mocks to intercept our API calls when we do not have an actual API-capable device but still want to run our tests.

> A few years ago, Matt Oswalt announced the **Testing On Demand: Distributed** (**ToDD**) validation tool for network changes. It is an open source framework aimed at testing network connectivity and distributed capacity. You can find more information about the project on its GitHub page: https://github.com/toddproject/todd. Oswalt also talked about the project on this Packet Pushers Priority Queue 81, Network Testing with ToDD: https://packetpushers.net/podcast/podcasts/pq-show-81-network-testing-todd/.

In this section, let's look at how we can write tests that are relevant to the networking world. There is no shortage of commercial products when it comes to network monitoring and testing. Over the years, I have come across many of them. However, in this section, I prefer to use simple, open source tools for our tests.

Testing for reachability

Often, the first step of troubleshooting is to conduct a small reachability test. For network engineers, `ping` is our best friend when it comes to network reachability tests. It is a way to test the reachability of a host on an IP network by sending a small package across the network to the destination.

We can automate the `ping` test via the `os` module or the `subprocess` module:

```
>>> import os
>>> host_list = ['www.cisco.com', 'www.google.com']
>>> for host in host_list:
```

Test-Driven Development for Networks

```
...         os.system('ping -c 1 ' + host)
...
PING e2867.dsca.akamaiedge.net (69.192.206.157) 56(84) bytes of data.
64 bytes from a69-192-206-157.deploy.static.akamaitechnologies.com
(69.192.206.157): icmp_seq=1 ttl=54 time=14.7 ms

--- e2867.dsca.akamaiedge.net ping statistics ---
1 packets transmitted, 1 received, 0% packet loss, time 0ms
rtt min/avg/max/mdev = 14.781/14.781/14.781/0.000 ms
0
PING www.google.com (172.217.3.196) 56(84) bytes of data.
64 bytes from sea15s12-in-f196.1e100.net (172.217.3.196): icmp_seq=1 ttl=54
time=12.8 ms

--- www.google.com ping statistics ---
1 packets transmitted, 1 received, 0% packet loss, time 0ms
rtt min/avg/max/mdev = 12.809/12.809/12.809/0.000 ms
0
>>>
```

The `subprocess` module offers the additional benefit of catching the output back:

```
>>> import subprocess
>>> for host in host_list:
...     print('host: ' + host)
...     p = subprocess.Popen(['ping', '-c', '1', host],
stdout=subprocess.PIPE)
...     print(p.communicate())
...
host: www.cisco.com
(b'PING e2867.dsca.akamaiedge.net (69.192.206.157) 56(84) bytes of
data.\n64 bytes from a69-192-206-157.deploy.static.akamaitechnologies.com
(69.192.206.157): icmp_seq=1 ttl=54 time=14.3 ms\n\n---
e2867.dsca.akamaiedge.net ping statistics ---\n1 packets transmitted, 1
received, 0% packet loss, time 0ms\nrtt min/avg/max/mdev =
14.317/14.317/14.317/0.000 ms\n', None)
host: www.google.com
(b'PING www.google.com (216.58.193.68) 56(84) bytes of data.\n64 bytes from
sea15s07-in-f68.1e100.net (216.58.193.68): icmp_seq=1 ttl=54 time=15.6
ms\n\n--- www.google.com ping statistics ---\n1 packets transmitted, 1
received, 0% packet loss, time 0ms\nrtt min/avg/max/mdev =
15.695/15.695/15.695/0.000 ms\n', None)
>>>
```

These two modules prove to be very useful in many situations. Any command we can execute in the Linux and Unix environment can be executed via the `os` or `subprocess` module.

Testing for network latency

The topic of network latency can sometimes be subjective. Working as a network engineer, we are often faced with the user saying that the network is slow. However, slow is a very subjective term. If we could construct tests that turn subjective terms into objective values, it would be very helpful. We should do this consistently so that we can compare the values over a time series of data.

This can sometimes be difficult to do since the network is stateless by design. Just because one packet is successful does not guarantee success for the next packet. The best approach I have seen over the years is just to use ping across many hosts frequently and log the data, conducting a ping-mesh graph. We can leverage the same tools we used in the previous example, catch the return-result time, and keep a record:

```
$ cat chapter13_10_ping.py
#!/usr/bin/env python3

import subprocess

host_list = ['www.cisco.com', 'www.google.com']

ping_time = []

for host in host_list:
    p = subprocess.Popen(['ping', '-c', '1', host], stdout=subprocess.PIPE)
    result = p.communicate()[0]
    host = result.split()[1]
    time = result.split()[14]
    ping_time.append((host, time))

print(ping_time)
```

In this case, the result is kept in a tuple and put into a list:

```
$ python3 chapter13_10_ping.py
[(b'e2867.dsca.akamaiedge.net', b'time=13.8'), (b'www.google.com', b'time=14.8')]
```

This is by no means perfect, and is merely a starting point for monitoring and troubleshooting. However, in the absence of other tools, this offers some baseline of objective values.

Testing for security

We already saw the best tool for security testing in Chapter 6, *Network Security with Python*, with Scapy, in my opinion. There are lots of open source tools for security, but none offers the flexibility that comes with constructing our packets.

Another great tool for network security testing is hping3 (http://www.hping.org/). It offers a simple way to generate a lot of packets at once. For example, you can use the following one-liner to generate a TCP Syn flood:

```
# DON'T DO THIS IN PRODUCTION #
echou@ubuntu:/var/log$ sudo hping3 -S -p 80 --flood 192.168.1.202
HPING 192.168.1.202 (eth0 192.168.1.202): S set, 40 headers + 0 data bytes
hping in flood mode, no replies will be shown
^C
--- 192.168.1.202 hping statistic ---
2281304 packets transmitted, 0 packets received, 100% packet loss
round-trip min/avg/max = 0.0/0.0/0.0 ms
echou@ubuntu:/var/log$
```

Again, since this is a command-line tool, we can use the subprocess module to automate any hping3 test we want.

Testing for transactions

The network is a crucial part of the infrastructure, but it is only a part of it. What the users care about is often the service that runs on top of the network. If the user is trying to watch a YouTube video or listen to a podcast but cannot, in their opinion, the service is broken. We might know that it is not the network transport, but that doesn't comfort the user.

For this reason, we should implement tests that are as similar to the user's experience as possible. In the example of a YouTube video, we might not be able to duplicate the YouTube experience 100% (unless you are part of Google), but we can implement a layer-seven service as close to the network edge as possible. We can then simulate the transaction from a client at a regular interval as a transactional test.

The Python HTTP standard library module is a module that I often use when I need to quickly test layer-seven reachability on a web service:

```
# Python 2
$ python -m SimpleHTTPServer 8080
Serving HTTP on 0.0.0.0 port 8080 ...
127.0.0.1 - - [25/Jul/2018 10:14:39] "GET / HTTP/1.1" 200 -
```

```
# Python 3
$ python3 -m http.server 8080
Serving HTTP on 0.0.0.0 port 8080 ...
127.0.0.1 - - [25/Jul/2018 10:15:23] "GET / HTTP/1.1" 200 -
```

If we can simulate a full transaction for the expected service, that is even better. But the Python simple `HTTP` server module in the standard library is always a great one for running some ad hoc web service tests.

Testing for network configuration

In my opinion, the best test for network configuration is using standardized templates to generate the configuration and back up the production configuration often. We have seen how we can use the Jinja2 template to standardize our configuration per device type or role. This will eliminate many of the mistakes caused by human error, such as copy and paste.

Once the configuration is generated, we can write tests against the configuration for known characteristics that we would expect before we push the configuration to production devices. For example, there should be no overlap of IP address in all of the network when it comes to loopback IP, so we can write a test to see whether the new configuration contains a loopback IP that is unique across our devices.

Testing for Ansible

For the time I have been using Ansible, I can not recall using a `unittest` like tool to test a Playbook. For the most part, the Playbooks are utilizing modules that were tested by the module developers.

Ansible provides unit tests for their library of modules. Unit tests in Ansible are currently the only way to drive tests from Python within Ansible's continuous-integration process. The unit tests that are run today can be found under `/test/units` (https://github.com/ansible/ansible/tree/devel/test/units).

The Ansible testing strategy can be found in the following documents:

- **Testing Ansible**: https://docs.ansible.com/ansible/2.5/dev_guide/testing.html
- **Unit tests**: https://docs.ansible.com/ansible/2.5/dev_guide/testing_units.html
- **Unit testing Ansible modules**: https://docs.ansible.com/ansible/2.5/dev_guide/testing_units_modules.html

One of the interesting Ansible testing frameworks is **molecule** (https://pypi.org/project/molecule/2.16.0/). It intends to aid in the development and testing of Ansible roles. Molecule provides support for testing with multiple instances, operating systems, and distributions. I have not used this tool, but it is where I would start if I wanted to perform more testing on my Ansible roles.

Pytest in Jenkins

Continuous-integration (**CI**) systems, such as Jenkins, are frequently used to launch tests after each of the code commits. This is one of the major benefits of using a CI system. Imagine that there is an invisible engineer who is always watching for any change in the network; upon detecting change, the engineer will faithfully test a bunch of functions to make sure that nothing breaks. Who wouldn't want that?

Let's look at an example of integrating `pytest` into the Jenkins tasks.

Jenkins integration

Before we can insert the test cases into our continuous integration, let's install some of the plugins that can help us visualize the operation. The two plugins we will install are **build-name-setter** and **Test Result Analyzer**:

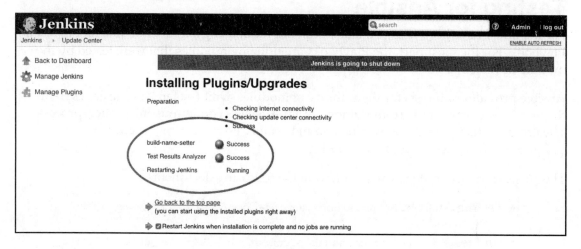

Jenkins plugin installation

The test we will run will reach out to the NXOS device and retrieve the operating system version number. This will ensure that we have API reachability to the Nexus device. The full script content can be read in `chapter13_9_pytest_4.py` the relevant `pytest` portion and result are as follows:

```
def test_transaction():
    assert nxos_version != False

## Test Output
$ pytest chapter13_9_pytest_4.py
============================ test session starts
===============================
platform linux -- Python 3.5.2, pytest-3.6.3, py-1.5.4, pluggy-0.6.0
rootdir: /home/echou/Chapter13, inifile:
collected 1 item

chapter13_9_pytest_4.py . [100%]

=========================== 1 passed in 0.13 seconds
===========================
```

We will use the `--junit-xml=results.xml` option to produce the file Jenkins needs:

```
$ pytest --junit-xml=results.xml chapter13_9_pytest_4.py
$ cat results.xml
<?xml version="1.0" encoding="utf-8"?><testsuite errors="0" failures="0" name="pytest" skips="0" tests="1" time="0.134"><testcase classname="chapter13_9_pytest_4" file="chapter13_9_pytest_4.py" line="25" name="test_transaction" time="0.0009090900421142578"></testcase></testsuite>
```

The next step would be to check this script into the GitHub repository. I prefer to put the test under its directory. Therefore, I created a `/test` directory and put the test file there:

Project repository

We will create a new project named `chapter13_example1`:

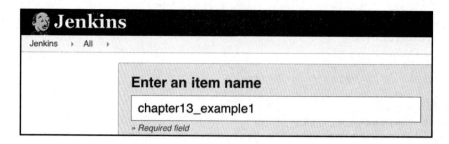

Chapter 13 example 1

We can copy over the previous task, so we do not need to repeat all the steps:

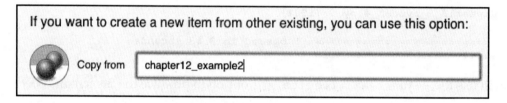

Copy task from chapter 12 example 2

In the execute shell section, we will add the `pytest` step:

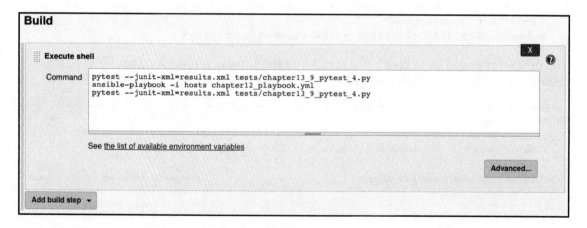

Project execute shell

We will add a post-build step of **Publish JUnit test result report**:

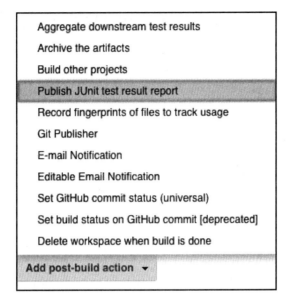

Post-build step

We will specify the `results.xml` file as the JUnit result file:

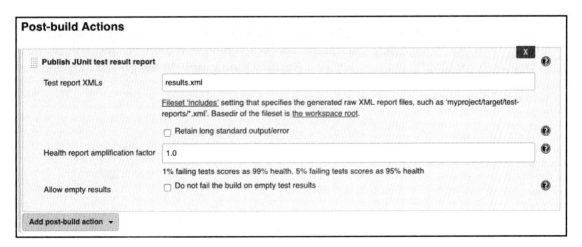

Test report XML location

After we run the build a few times, we will be able to see the **Test Result Analyzer** graph:

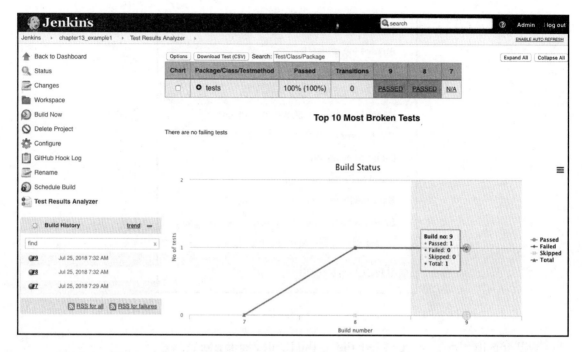

Test result analyzer

The test result can also be seen on the project homepage. Let's introduce a test failure by shutting down the management interface of the Nexus device. If there is a test failure, we will be able to see it right away on the **Test Result Trend** graph on the project dashboard:

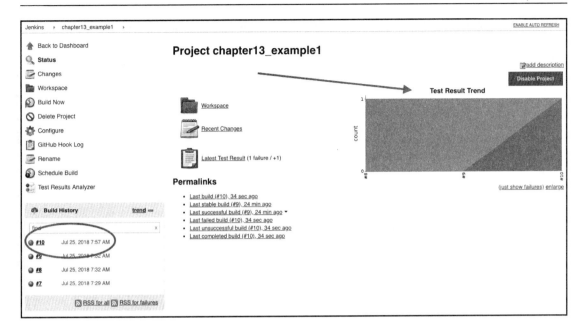

Test result trend

This is a simple but complete example. There are many ways we can integrate testing into Jenkins.

Summary

In this chapter, we looked at test-driven development and how it can be applied to network engineering. We started with an overview of TDD; then we looked at examples of using the `unittest` and `pytest` Python modules. Python and simple Linux command-line tools can be used to construct various tests for network reachability, configuration, and security.

We also looked at how we can utilize testing in Jenkins, a continuous-integration tool. By integrating tests into our CI tool, we can gain more confidence in the sanity of our change. At the very least, we hope to catch any errors before our users do.

Simply put, if it is not tested, it is not trusted. Everything in our network should be programmatically tested as much as possible. As with many software concepts, test-driven development is a never-ending service wheel. We strive to have as much test coverage as possible, but even at 100% test coverage, we can always find new ways and test cases to implement. This is especially true in networking, where the network is often the internet, and 100% test coverage of the internet is just not possible.

Other Books You May Enjoy

If you enjoyed this book, you may be interested in these other books by Packt:

Practical Network Automation
Abhishek Ratan

ISBN: 9781788299466

- Get the detailed analysis of Network automation
- Trigger automations through available data factors
- Improve data center robustness and security through specific access and data digging
- Get an Access to APIs from Excel for dynamic reporting
- Set up a communication with SSH-based devices using netmiko
- Make full use of practical use cases and best practices to get accustomed with the various aspects of network automation

Other Books You May Enjoy

Network Analysis using Wireshark 2 Cookbook - Second Edition
Nagendra Kumar Nainar, Yogesh Ramdoss, Yoram Orzach

ISBN: 9781786461674

- Configure Wireshark 2 for effective network analysis and troubleshooting
- Set up various display and capture filters
- Understand networking layers, including IPv4 and IPv6 analysis
- Explore performance issues in TCP/IP
- Get to know about Wi-Fi testing and how to resolve problems related to wireless LANs
- Get information about network phenomena, events, and errors
- Locate faults in detecting security failures and breaches in networks

Other Books You May Enjoy

Practical AWS Networking
Mitesh Soni

ISBN: 9781788398299

- Overview of all networking services available in AWS.
- Gain Work with load balance application across different regions.
- Learn auto scale instance based on the increase and decrease of the traffic.
- Deploy application in highly available and fault tolerant manner.
- Configure Route 53 for a web application.
- Troubleshooting tips and best practices at the end

Python Network Programming Cookbook - Second Edition
Pradeeban Kathiravelu, Dr. M. O. Faruque Sarker

ISBN: 9781786463999

- Develop TCP/IP networking client/server applications
- Administer local machines' IPv4/IPv6 network interfaces
- Write multi-purpose efficient web clients for HTTP and HTTPS protocols
- Perform remote system administration tasks over Telnet and SSH connections
- Interact with popular websites via web services such as XML-RPC, SOAP, and REST APIs
- Monitor and analyze major common network security vulnerabilities
- Develop Software-Defined Networks with Ryu, OpenDaylight, Floodlight, ONOS, and POX Controllers
- Emulate simple and complex networks with Mininet and its extensions for network and systems emulations
- Learn to configure and build network systems and Virtual Network Functions (VNF) in heterogeneous deployment environments
- Explore various Python modules to program the Internet

Leave a review - let other readers know what you think

Please share your thoughts on this book with others by leaving a review on the site that you bought it from. If you purchased the book from Amazon, please leave us an honest review on this book's Amazon page. This is vital so that other potential readers can see and use your unbiased opinion to make purchasing decisions, we can understand what our customers think about our products, and our authors can see your feedback on the title that they have worked with Packt to create. It will only take a few minutes of your time, but is valuable to other potential customers, our authors, and Packt. Thank you!

Index

A

access control list
  reference 201
Access Control Lists (ACLs) 341
access lists
  about 196
  implementing, with Ansible 197, 199
  MAC access lists 200
action plugin
  reference 176
Amazon Elastic Compute Cloud (EC2)
  about 328
  reference 328
Amazon Elasticsearch Service
  reference 279
Amazon GuardDuty
  about 349
  reference 349
Amazon Resource Names (ARNs)
  about 328
  reference 328
Amazon Virtual Private Cloud (Amazon VPC) 328
Amazon Web Services (AWS)
  about 317
  reference 317, 318, 320
  setting up 318, 320
Ansible Galaxy
  about 174
  reference 174
Ansible Jinja2 template
  reference 133
Ansible playbook
  about 118, 120, 121, 122
  inventory file 119, 120
  public key authorization 118
Ansible UFW module
  reference 206
Ansible Vault
  about 167, 168, 169
  reference 167, 168
Ansible, advantages
  about 122
  agentless 122, 123
  extensible 124
  idempotent 123
  network vendor support 125, 126
  simplification 124
Ansible
  about 113, 114
  advantages 197
  architecture 126
  Arista example 143
  Cisco example 136
  conditional statements 148
  connection example 139
  control node installation 115
  different versions, executing from source 116, 117
  example 114
  include statement 169, 170
  Juniper example 141
  loops 155
  networking modules 133
  reference 22, 114, 115, 139
  roles 169, 171, 172, 173
  setting up 117
API structured output
  versus screen scraping 74, 75, 77
AppleTalk 15
Application Program Interface (API) 37, 71
architecture, Ansible
  about 126
  inventory 126, 128, 129

templates 126
templates, with Jinja2 133
variables 126, 129, 132, 133
YAML 126, 127
Arista eAPI management
  about 98
  eAPI, preparation 99, 100, 101
  examples 102, 103
  reference 102
Arista Networks
  about 98
  reference 98
Arista Pyeapi library
  about 104
  examples 105, 106, 107
  installation 104
  reference 104, 105, 107
Arista Python API
  about 98
  Arista eAPI management 98
  Arista Pyeapi library 104
Arista vEOS
  reference 49
automation
  bad automation 69
Availability Zones (AZ) 324
AWS CLI
  Python SDK 321, 322, 323
  reference 321
AWS CloudFront 327
AWS Direct Connect 327
AWS Direct Connect locations
  reference 346
AWS Edge locations 327
AWS Global Infrastructure
  reference 324
AWS network
  overview 324, 325, 326, 327, 328
  services 349
AWS Shield
  about 349
  reference 349
AWS Transit Centers 327
AWS Transit VPC
  about 349
  reference 349
AWS WAF
  about 349
  reference 349

B

Beats 278
Boto3 VPC API
  reference 333
Boto3
  reference 323
Bring Your Own Device (BYOD) 193
built-in types, Python
  about 24
  mapping 28
  None type 25
  numerics 25
  sequences 25, 27, 28
  sets 29

C

Cacti
  about 236
  installation 237, 238, 239
  Python script, as input source 239, 240
  reference 236, 240
change-advisory board (CAB) 382
Cisco API
  about 78
  Cisco NX-API 79
  YANG models 85
Cisco Application Centric Infrastructure (ACI)
  about 78, 86, 87, 88
  reference 86
Cisco Certified Internetwork Expert (CCIE) 41
Cisco Connection Online (CCO) 46
Cisco dCloud
  reference 42
Cisco DevNet
  about 46, 47
  reference 42, 46
Cisco IOS MIB locator
  reference 213
Cisco NX-API
  about 79

device preparation 79, 80
examples 80, 81, 82, 83, 84, 85
installation 79, 80
reference 79
Cisco VIRL
about 41
advantages 42
reference 43
tips 43, 44, 45
classes, Python 33
client-server model 15
cloud computing 317
cloud data centers 11, 12
CloudFormation
about 335
for automation 335, 336, 338, 339
reference 335
CloudFront CDN services 348
Command Line Interface (CLI)
about 37
challenges 38, 39
communication protocols
reference 134
conditional statements, Ansible
about 148
network facts 151, 152, 153
network module conditional 153, 155
reference 150
when clause 148, 149, 150
configuration backup
automating 375, 377, 378
Content Delivery Network (CDN) 348
Continuous Integration (CI)
about 383, 428
workflow 383
continuous integration
for networking 407
custom module
reference 175, 177
writing 174, 175, 176, 177, 178

D

Data Center Networking (DCN) 38
data centers
about 10
cloud data centers 11, 12
edge data centers 12
enterprise data centers 10
data model
about 77
reference 77
data modeling
for infrastructure as code 77, 78
data visualization
Matplotlib 222
Pygal 230
with Python 221
dCloud
about 46, 47
reference 47
declarative framework 113, 114
device ID API 305, 306
devices API 303, 304, 305
dictionary 28
Direct Connect
about 344, 345, 346
reference 346
Distributed Denial of Service (DDoS)
about 349
reference 206
Django Database
reference 288
Django
about 288
reference 288
DjangoCon 287
DOT format
reference 245
Dynamip 41

E

edge data centers 12
Elastic Cloud
reference 279
Elastic IP (EIP)
about 341
reference 342
Elasticsearch (ELK stack)
about 278, 279
hosted ELK service, setting up 279, 280

logstash format 280
Elasticsearch
  reference 278
Emulated Virtual Environment Next Generation (EVE-NG)
  reference 49
enterprise data centers 10
Equinix Cloud Exchange
  reference 346
Extensible Markup Language (XML) 89

F

Flask-SQLAlchemy
  about 298, 299
  reference 298
Flask
  about 288, 291
  additional resources 314
  HTTPié client 293, 294
  jsonify return 297
  reference 288, 291, 315
  setting up 289, 290, 291
  URL, generation 296
  URL, routing 294
  URL, variables 295
floor division 29
flow-based monitoring
  about 260
  IPFIX 260
  NetFlow 260
  ntop traffic monitoring 266, 268, 269
  sFlow 261, 274
FlowSet
  reference 265
functions, Python 32

G

Git, terminology
  branch 354
  checkout 354
  commit 354
  fetch 354
  merge 354
  pull 354
  ref 354
  repository 354
  tag 354
Git, with Python
  about 372
  GitPython 372
  PyGitHub 373, 375
Git
  about 352
  benefits 353
  collaboration technology 378
  examples 357, 358, 359, 361, 362, 363
  reference 354
  setting up 355
  software-development collaboration 378
  terminology 354
GitHub
  about 355
  collaborating, with pull requests 369, 370, 371, 372
  example 364, 365, 366, 367, 368
  reference 354
Gitignore
  about 356
  reference 357
GitPython
  reference 372
Global Information Tracker (GIT) 353
GNS3 48, 49
Grafana 278
Graphix
  installing 247
Graphviz
  about 245
  examples 248, 249, 250
  installing 247
  LLDP neighbor graphing 251, 252
  reference 248, 251
  setting up 246, 247
group variables
  about 164, 165, 167
  reference 165

H

host variables
  about 164, 166

reference 165
hosted ELK service
  setting up 279, 280
hosts 9
hping3
  reference 426
HTTP-bin
  reference 293
HTTPie client
  about 293, 294
  reference 293
HTTPie
  reference 294

I

idempotence
  reference 123
idempotency 68, 69
Identify and Access Management (IAM)
  about 327
  reference 327
IETF
  reference 260
include statement, Ansible 169, 170
infrastructure as code
  about 72
  data modeling 77, 78
  Intent-Driven Networking 73
  screen scraping, versus API structured output 74, 75, 77
Infrastructure-as-a-Service (IaaS)
  reference 317
InMon
  reference 274
Intent-Based Networking 73
Intent-Driven Networking 73
Intermediate Distribution Frame (IDF) 10
International Organization for Standardization (ISO) 13
International Telecommunication Union (ITU-T) 13
Internet Assigned Numbers Authority (IANA) 16
Internet Control Message Protocol (ICMP) 187
Internet of Things (IoT) 9
Internet Protocol (IP)
  about 14, 18, 19

Network Address Translation (NAT) 19
  routing 20
  security 19
Internet Service Provider (ISP) 9
internet
  data centers 10
  hosts 9
  network components 9
  overview 8, 9
  servers 9
inventory file
  about 119
  reference 119
inventory, Ansible
  about 128, 129
  reference 129
ios_command module
  reference 134
IPFIX 260

J

Jenkins
  example 387
  installing 384, 386, 387
  job, building for Python script 388, 389, 390, 392, 393, 394
  network continuous integration example 396, 397, 398, 401, 402, 403, 404, 405
  plugins, managing 394, 395, 396
  Pytest 428
  pytest, integrating 428, 429, 430, 431, 432, 433
  reference 384
  with Python 405, 407
Jinja2 template
  about 160
  reference 159
Jinja2
  about 133
  conditional statement 162, 163, 164
  loops 161
  reference 133
  templates 133
JSON-RPC 98
jsonrpclib
  reference 101

Juniper networks
   Network Configuration Protocol (NETCONF) 89
   PyEZ 93
   Python API 88
Juniper Olive 90
Juniper vMX
   reference 49

K

Kibana 278

L

lambda expressions
   reference 194
large data center 98
Link Layer Discovery Protocol (LLDP) 245
LLDP neighbor graphing
   about 251, 252
   final playbook 258, 259
   information retrieval 253
   Python parser script 254, 255, 257
Local Area Network (LAN) 9, 15
Logstash 278
logstash format
   reference 280
loops, Ansible
   about 155
   over dictionaries 156, 157, 158
   reference 158
   standard loops 155, 156

M

MAC access lists 200
Main Distribution Frame (MDF) 10
Management Information Base (MIB) 212
Matplotlib
   about 222
   example 222, 223, 224
   for SNMP results 225, 226, 227, 229
   installation 222
   reference 222, 225, 230
   resources 230
modules, Ansible
   reference 119
modules, Python 34, 35

molecule
   about 428
   reference 428
Multi Router Traffic Grapher (MRTG)
   reference 236
Multiprotocol Label Switching (MPLS) 88

N

NAPALM
   reference 109
NAT Gateway
   about 327, 342, 344
   using 343
ncclient library
   about 80
   reference 80
NetFlow
   about 260
   parsing, with Python 261
   Python socket 263, 264
   reference 263
   struct 263, 264
Netmiko
   reference 63, 109
Network Address Translation (NAT) 19
network automation
   reference 407
network components 9
Network Configuration Protocol (NETCONF)
   about 89
   characteristics 89
   device preparation 89, 90
   examples 91, 93
   reference 89
network dynamic operations
   about 306, 307, 308
   asynchronous operations 309, 310, 311
network module conditional 153, 155
network modules
   reference 125
network protocol suites
   about 15
   Internet Protocol (IP) 18, 19
   Transmission Control Protocol (TCP) 16
   User Datagram Protocol (UDP) 17, 18

network resource API
  about 298
  device ID API 305, 306
  devices API 303, 304, 305
  Flask-SQLAlchemy 298, 299
  network content API 300, 301, 302
network scaling services
  about 347
  CloudFront CDN services 348
  Elastic Load Balancing (ELB) 347
  Route53 DNS service 348
network security tools
  private VLANs 204
  reference 206
  UFW, with Python 205, 206
network security
  setting up 182, 183, 184, 185
networking modules, Ansible
  about 133
  facts 133, 134
  local connections 133, 134
  provider arguments 134, 135, 136
networking
  continuous integration 407
  testing, for Ansible 427
  testing, for network configuration 427
  testing, for network latency 425
  testing, for reachability 423, 424
  testing, for security 426
  testing, for transactions 426
  tests, writing 423
ntop
  about 266, 268, 269
  Python extension 270, 271, 272, 273
NumPy
  about 222
  reference 222
nxso_snmp_contact module
  reference 130

O

Object Identifier (OID) 212
object-oriented programming (OOP) 33
Open Shortest Path First (OSPF) 183
Open System Interconnection (OSI) model
  about 13
  Application layer 14
  Data link layer 14
  Physical layer 14
Organizationally Unique Identifier (OUI) 200

P

Paramiko library
  about 59
  drawbacks 68
  features 65
  for servers 65, 66
  implementing 66, 67, 68
  installation 60
  overview 61, 62
  program 64
  reference 60, 82
Pexpect library
  about 49
  drawbacks 68
  features 56, 59
  implementing 58
  installation 49, 50
  overview 50, 51, 53, 54
  program 55, 56
  reference 49, 57
  SSH 57
ping module
  reference 119
Ping of Death
  reference 194
Platform-as-a-Services (PaaS)
  reference 317
port address translation (PAT) 343
private VLANs
  about 205
  Community (C) port 205
  Isolated (I) port 205
  Promiscuous (P) port 204
PyEZ
  about 93
  examples 96, 97
  installation 94, 95, 96
  preparation 94, 95, 96
  reference 93, 94

Pygal
  about 230
  example 231, 232
  for SNMP results 233, 234, 235
  installation 230
  reference 230, 231
  resources 236
PyGitHub
  reference 373
PySNMP
  about 215, 216, 217, 219, 220, 221
  reference 215, 216
Pytest
  about 420
  example 420, 421, 423
  in Jenkins 428
  reference 420
Python API
  for Juniper networks 88
Python helper script
  for Logstash formatting 281, 282, 283
Python SDK 321, 322, 323
Python socket
  reference 263
Python web frameworks
  comparing 287, 288, 289
  reference 287
Python's unittest module
  about 416, 417, 419
  reference 419
Python
  built-in types 24
  classes 33
  control flow tools 30, 32
  executing 23, 24
  for Cacti 236
  for data visualization 221
  functions 32
  modules 34, 35
  NetFlow, parsing 261
  operating system 23
  operators 29
  overview 20, 21
  packages 34, 35
  reference 8, 21
  versions 22
  with Graphviz, examples 250

Q
Quora
  reference 288

R
Raspberry Pi file
  reference 309
Reddit
  reference 288
regular expressions
  reference 54
Remote Procedure Call (RPC) 98
Representational state transfer (REST)
  about 285
  reference 285
Requests
  reference 83
roles, Ansible
  about 171, 172, 173
  reference 171, 173
Round-Robin Database Tool (RRDtool)
  reference 236

S
Scalable Vector Graphics (SVG) 230
Scapy
  about 185
  common attacks 194, 195
  examples 187, 188
  installing 186
  ping collection 193
  reference 185, 195
  resources 195
  sniffing 188
  TCP port scan 190, 192
screen scraping
  drawbacks 75
  versus API structured output 74, 75, 77
security 311
Service-Level Agreement (SLA) 345
sets 29
sFlow 261, 274

sFlow-RT
  reference 276
  with Python 274, 275, 276, 277
SFlowtool
  about 274, 275, 276, 277
  reference 275
sflowtool
  reference 274
Simple Network Management Protocol (SNMP)
  about 211, 212
  PySNMP 215, 216, 217, 219, 220, 221
  setting up 210, 212, 214
SNMP Object Navigator
  reference 214
Software Defined Networking (SDN) 317
Software-as-a-Service (SaaS)
  reference 317
SSH
  Pexpect library 57
struct
  reference 263
Supervisor
  reference 314
Syslog search
  about 201
  reference 201
  with RE module 202, 204

T

Telecommunication Standardization Sector 13
template module
  reference 158
templates
  about 158, 159, 160
  Jinja2 template 160
  with Jinja2 133
Ternary Content-Addressable Memory (TCAM) 196
test-driven development (TDD)
  about 409
  definitions 411
  overview 410
Testing On Demand
  Distributed (ToDD) 423
testing strategy

  reference 427
topology, as code
  about 411, 413, 415, 416
  Python's unittest module 416, 417, 419
traditional change-management process 381, 383
Transmission Control Protocol (TCP)
  about 14, 16
  characteristics 16
  data transfer 17
  functions 16
  messages 17
  reference 17

U

Uncomplicated Firewall (UFW) 201
unittest module
  reference 418
User Datagram Protocol (UDP) 17

V

variables, Ansible
  about 129, 132, 133
  reference 130, 131
vendor-neutral libraries 109
VIRL on Packet
  reference 42
virlutils
  reference 42
Virtual Internet Routing Lab (VIRL)
  reference 41
virtual lab
  advantages 40
  Cisco DevNet 46, 47
  Cisco VIRL 41, 42
  constructing 39, 40, 41
  dCloud 46, 47
  disadvantages 40
  GNS3 48, 49
Virtual Local Area Networks (VLANs) 204
Virtual Private Cloud (VPC) 327
virtual private cloud
  about 328
  automation, with CloudFormation 335
  creating 329, 330, 331, 332, 333
  Elastic IP (EIP) 341

NAT Gateway 342, 344
network ACL 339, 340, 341
route tables 333, 334
route targets 333, 334
security groups 339, 340, 341
virtual private clouds
  automation, with CloudFormation 335, 336, 338, 339
Virtual Private Gateway 345
virtualenv
  reference 289
VPC peering
  reference 339
VPN Gateway 344
vSRX
  reference 49

W

when clause 148, 149, 150

X

X-Pack 278

Y

YAML
  about 127
  reference 120
YANG models
  and Cisco API 85
  reference 86
Yet Another Next Generation (YANG) 78